T0289545

QGIS for Ecologists

QGIS for Ecologists

*An Introduction to Mapping
for Ecological Surveys*

Stephanie Kim Miles

DATA IN THE WILD

Pelagic Publishing | www.pelagicpublishing.com

First published in 2024 by
Pelagic Publishing
20–22 Wenlock Road
London N1 7GU, UK

www.pelagicpublishing.com

QGIS for Ecologists: An Introduction to Mapping for Ecological Surveys

Copyright © 2024 Stephanie Kim Miles

The moral rights of the author have been
asserted by them in accordance with the
Copyright, Designs and Patents Act 1988.

All rights reserved. Apart from short excerpts for use
in research or for reviews, no part of this document
may be printed or reproduced, stored in a retrieval
system, or transmitted in any form or by any means,
electronic, mechanical, photocopying, recording,
now known or hereafter invented or otherwise
without prior permission from the publisher.

https://doi.org/10.53061/IHQS7086

British Library Cataloguing in Publication Data
A catalogue record for this book is available from the British Library

ISBN 978-1-78427-297-5 Hbk
ISBN 978-1-78427-298-2 Pbk
ISBN 978-1-78427-299-9 ePub
ISBN 978-1-78427-300-2 PDF

Cover image: example of an advanced habitats map

Typeset by S4Carlisle Publishing Services, Chennai, India

Contents

Part I

Fundamentals

1. Introduction

1.1   About this book

This book is designed to teach you the basic stages of mapping for ecological projects. It uses QGIS, an open-source system as this is the most accessible platform for all ecologists to produce maps for reports.

This book guides you, the beginner mapmaker, through production of maps for the day-to-day projects of ecologists working in ecological consultancy. If you do not know where to start or how to use QGIS, this is the practical guide for you. There is no jargon, just what you need to know and what you need to do to create the maps and to get the information from them you need for reporting.

This book includes web link to downloadable data for use in producing the maps described. You will learn how to produce several different types of maps for ecological reports, including a basic survey map, an aerial imagery survey map, a designated sites map, a desk-study map, a protected species map and a Phase 1 habitat map. As a part of this you will also learn how to download third-party maps and datasets, georeference images, import GPX files into QGIS.

The book includes workflows that you can follow with your own data.

1.2   What you will learn

Basic maps

The basic maps chapter introduces how to use QGIS and the types of data needed to make maps. By the end of this chapter you will learn:

- The difference between vector and raster data
- How to produce a basic map
- Where to download basemaps

Survey maps

The survey map chapter shows you how to create a basic map to be taken into the field for your initial site visit.

You will learn:

- How to use online basemaps
- What is a plugin?
- How to produce a survey map for field use
- What are Coordinate Reference Systems?
- How to produce an aerial imagery map for field use

Designated sites maps

The designated sites chapter shows you how to create a map from data previously collected on the site by other organizations or surveys.

You will learn:

- How to set up a Project
- How to use Coordinate Reference Systems
- How to import existing vector data
- How to perform basic analysis on vector data
- How to produce a designated study map from existing data
- Where to download designated sites data

Desk-study maps

The desk-study map chapter shows you how to map data from a spreadsheet of records from protected species survey.

You will learn:

- How to create and import point vector data from a spreadsheet
- How to import data from NBN
- How to troubleshoot in data creation
- How to produce desk species maps from created data

Protected species survey maps

The Protected Species Survey maps chapter shows you how to map data from a field survey on signs of protected species.

You will learn:

- How to create and import point vector data from GPS field survey
- How to troubleshoot in data creation
- How to produce a protected species map from created data

Georeferencing maps

The Georeferencing maps chapter shows you how to georeference scanned field maps.

You will learn:

- How to prepare field maps for use in QGIS
- How to georeference a field map with grid-lines
- How to georeference a field map without grid-lines

Habitat survey maps

The Habitat Survey maps chapter shows you how to map data from a field surveys of broad habitat types.

You will learn:

- How to create and edit points, lines and polygons
- How to troubleshoot in data creation and perform analysis on created data
- How to produce a Phase 1 habitat map from created data

1.3   How to use this book

Part 1: Fundamentals

You are in Chapter 1, which provides an overview of the book and a summary of what you will learn in its pages.

Chapter 2 offers an introduction to the author, their background and how they came to write the book.

Chapter 3 explains the brief of your ecological project and why you need QGIS for it.

Part 2: Workbook

Chapter 4 explains how to download QGIS, install QGIS and download supplementary data to use for the exercises contained in Chapters 5–11.

Chapters 5–11 contain workflows and screenshots of how to produce maps for ecological consultancy projects. You can follow these by downloading the data and unzipping the folder to your computer. Alternatively, you can use your own data and follow the same workflows. Within the workflows are exercises designed to increase your understanding by asking you to stop and assess your learning. They also show you how to discover the information you need for reporting as you produce maps to insert into reports.

Part 3: Answers to Exercises, Chapter 11

Chapter 12 contains the answers to the exercises in Chapters 5–11.

For ease of reference once you have completed the book, Part 4: Workflows, Chapter 12 provides workflows without the screenshots and exercises, so you can easily repeat the methodology you have learnt.

In the Appendix is a glossary of terms, as despite the promise of no jargon, there are a few technical terms that are unavoidable.

It is recommended that you use a mouse while working in QGIS. This makes life a lot easier in navigating the interface and when drawing features. Navigating the interface involves moving the cursor across the screen, scrolling in and out of the map, left- and right-clicking on icons, clicking and dragging to draw, and clicking once more to finish drawing. If you wish to proceed without a mouse, do so at your own risk! The main danger is increased frustration, but in my opinion it is not worth attempting to use QGIS without one. I use a thumb-roller mouse, which enables the cursor to be moved by moving the thumb, while keeping the wrist still.

It is recommended that you use a separate screen and increase font size while working in QGIS.

This is especially the case if you are working on a laptop. The QGIS interface contains a lot of small text and icons. To improve this, you can increase the default font size in your operating system settings.

It is recommended that you take frequent breaks while working in QGIS. Computer use is associated with a number of health issues due to sitting and screen use.[1] Computer mapping is especially intense on the eyes, particularly when digitizing data.

---

1   For more information on safe computer use, consult the Health and Safety Executive Working safely with display screen equipment, current web address: https://www.hse. gov.uk/msd/dse/

2. About the author

Stephanie is GIS Officer for the Bumblebee Conservation Trust. They have over 10 years of experience working with ecological and environmental GIS data for consultancies, conservation projects and the public sector. Stephanie has a BSc in Environmental Science from Lancaster University and a Masters in Conservation and Biodiversity from the University of Exeter.

"I wanted to write this book to help fellow ecologists to use QGIS. Learning GIS can be challenging, so I hope this step-by-step workbook can ease this difficultly."

2.1 Acknowledgements

Thank you to:

My first GIS teachers Stuart Gibson and Georgina Chapman, without whom I would not have started in this field.

Past and present students and colleagues who have taught me a lot about teaching and providing GIS support in ecology and conservation.

My friends and family for initial comments and edits of my proposal and a lot of support.

Paul Loose, originally the coauthor of this book, for editing and support in the early stages.

Hugo Gault for GIS testing and editing in the final stages.

Nigel Massen, David Hawkins and rest of the team at Pelagic Publishing.

3. Why QGIS?

If you are working or hope to work in ecological consultancy or conservation you will be working on a number of different sites. For survey and reporting on these sites you will need to produce maps. QGIS is a good choice of software for this because it is free and open source (and there is this handy guide you have bought to help you learn how to use it!).

To keep things interesting, we will be looking at different sites and different sorts of data. In reality, for a project you would need all or most of the types of maps we will work through as part of ecological site assessment.

For the first site you need a basic map of the site and a survey map to take out into the field. For the second site you need a designated site map of protected areas that could be impacted by any works on the site. For the third and fourth sites you need a desk-study map from existing data. You have undertaken a protected species survey at the fifth site and returned with GPS data and want to create a protected species map of the field signs you recorded on the survey. For the last map, we return to the first site and your aerial image which you have hand drawn annotations in the field that you intend to use to produce a Phase 1/UK habitats map of the broad habitats you recorded.

For these sites you are provided with materials (see page 14). The basemaps, aerials, environmental designation and species records datasets provided are real-world data. These maps and datasets you can download, map and modify commercially except where explicitly stated otherwise, for instance the aerial images used here cannot be used as basemaps for printing in reports but can be used for digitization. It is recommended that you first follow the exercises with the data provided then use your own data as per the instructions at the end of each chapter and then return to the workbook a second time to work through again with your own data using the workflows.

Part II

Workbook

Please note that the data required for use with
Part II: Workbook can be downloaded here
(see page 14):
https://pelagicpublishing.com/pages/
qgis-for-ecologists-online-resources

4. How to download QGIS and supplementary data

4.1 How to download to and install QGIS on your computer

You will learn:

- The process of downloading QGIS on Windows or Mac.
- Where to install GIS on your computer.

I recommend that you use QGIS on a desktop computer with a mouse. Laptops can run QGIS but are slower and prone to crashing while working with large datasets.

In your web browser go to the QGIS download page, which is currently: https://www.qgis.org/en/site/forusers/download.html

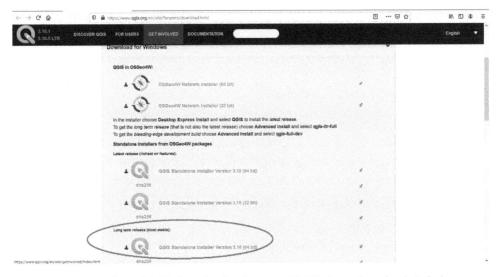

Figure 4.1 Screenshot of QGIS download webpage with Windows download circled

How to find out whether you have a 64 bit or 32 bit system.

Go to '**Control Panel\System and Security\System**' information can be found under System>System type (note QGIS is phasing out 32 bit support).

Scroll down to **Download for Windows**

Choose '**Long term release (e.g. for corporate users)**' this is the most stable version

Click on: **QGIS Standalone Installer Version 3.xx (64 bit)**

Note: xx is used here where you would see two numbers after the point for the version, QGIS is frequently updated, so version numbers change regularly.

Pop-up box will appear, click: '**Save File**'

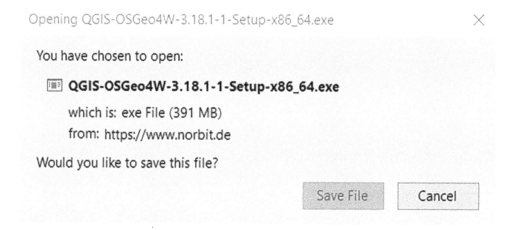

Figure 4.2 Save file Windows pop-up box

Open the installation file from Windows Explorer and follow the installation wizard.

Install QGIS to C̲: drive if possible, as GIS systems usually run best on C: drive. QGIS defaults to installation on C: drive, so you can just accept where it chooses to install.

You do not need to install any of the additional datasets.

QGIS will take a bit of time to install and then will ask to restart your computer. If not prompted to restart, do so before using QGIS.

For **Mac** users:

Scroll down to **Download for Mac.**

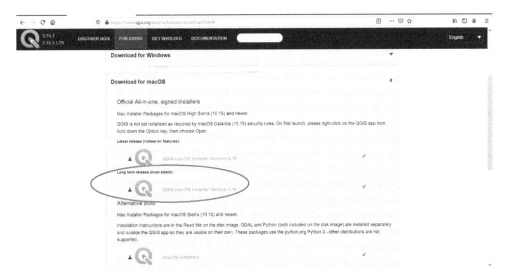

Figure 4.3 Screenshot of QGIS download webpage with Mac download circled

Click on: **QGIS macOS Installer version 3.xx**

Pop-up box will appear, select **Save File** and click: **OK**

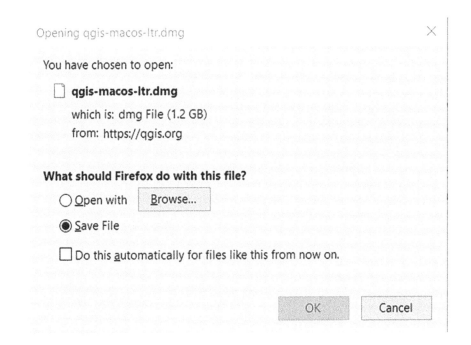

Figure 4.4 Save file mac pop-up box

This will download a zip file which you will need to extract.

To do so you will need to **override the security settings**, which will prompt a query about QGIS being an "unidentified developer":

Right-click the installer file and select **Open**, this will trigger an extra option in the security warning select to install it anyway.

Install files in numerical order, they are labelled 1–4.

For users of other operating systems, please click on the appropriate tab follow the instructions for installation.

4.2 How to download supplementary materials

In order to follow this workbook, supplementary materials have been provided at: https://pelagicpublishing.com/pages/qgis-for-ecologists-online-resources

Simply complete the form on the publisher's website in order to access these resources.

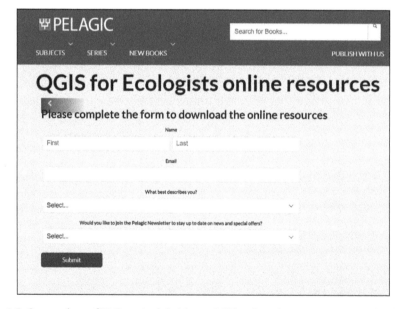

Figure 4.5 Screenshot of 'A Practical Guide to QGIS – Supplementary Materials' download webpage

Save to an appropriate location on your computer by creating a new folder for these materials in a location you will remember. This does not need to be C: drive if using a PC, for laptops with only a C: drive, save to C: drive.

Once the materials have downloaded, **right-click** on the folder and **left-click** on **Extract** to unzip the containing folder.

Great! You are now ready to begin mapping!

5. Basic maps

In order to make maps you first need to understand how to work with QGIS and the data it uses. By the end of this chapter you will learn:

- The difference between vector and raster data.
- How to import existing raster and vector data.
- How to produce a basic map for field use.
- Where to download basemaps.
- How to use Coordinate Reference Systems.

5.1 The QGIS Interface

To get us started we are going to open QGIS.

In the QGIS folder created on your desktop by the installation process you will have a shortcut called 'QGIS Desktop 3.xx', **double-click** on this to open. Alternatively, search for QGIS in the start menu. This will take a minute or two to load.

Figure 5.1 QGIS Opening screen

Once loaded you will be confronted with a rather complex screen. What you are looking for is the '**New Project**' button in the top left corner, **left-click** on this to open.

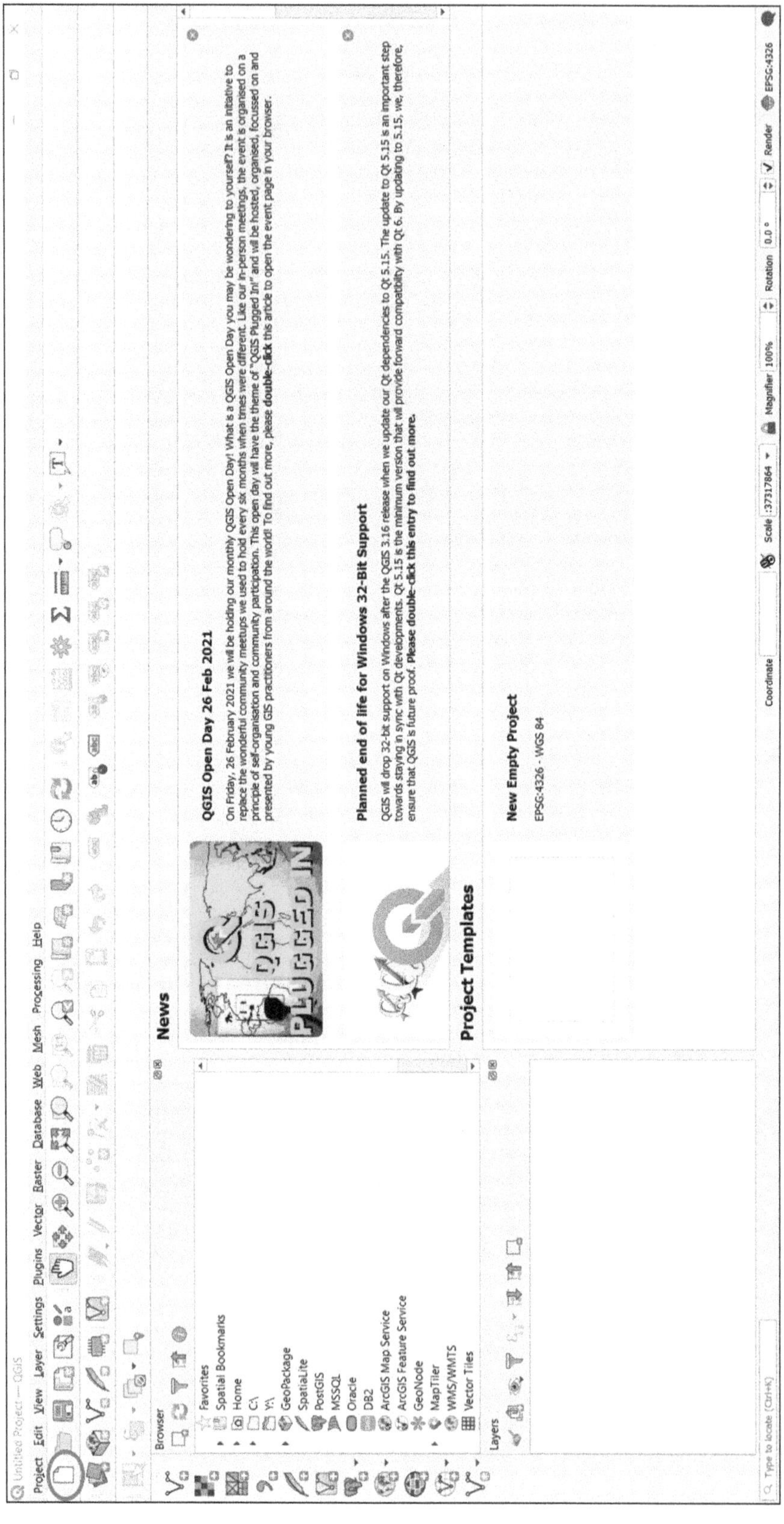

Figure 5.2 QGIS Opening screen with New Project button circled

Figure 5.3 QGIS Project Interface with 4 areas highlighted

Once open, take a look at the screen, this is the interface that lets us use QGIS bring in data and make maps.

There are four areas of the QGIS screen in front of us: the top menu and toolbars, the left panel containing the shortcut bar, browser panel and layers panel, the blank space where our map will go and the bottom info bar.

Annotate the screenshot above to help familiarize yourself with the interface. In the text, instructions will be provided using the following: top menu, top toolbar, shortcut bar, browser panel, layers panel, map area, bottom info bar.

With the mouse, hover over the buttons in the top toolbars

What do you notice?

....

Your panels and top toolbar may look different to the figure above; this is because it contains panels and multiple toolbars we could have turned on. To turn the panels and toolbars on and off **right-click** on the top toolbar. A pop-up tick box will appear. Experiment with turning layers on and off by ticking the boxes on and off.

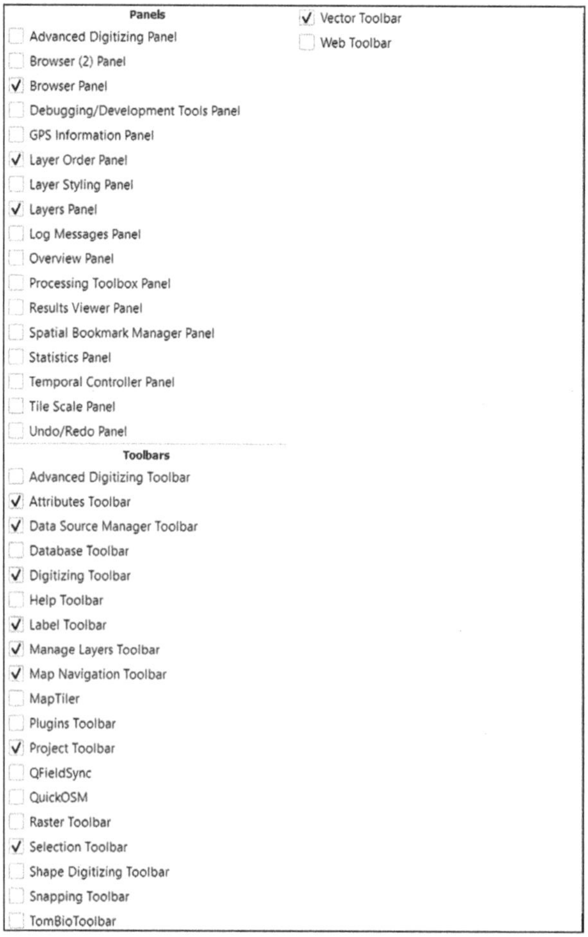

Figure 5.4 Panels and Toolbars pop-up tick box

What does the button below do and where can it be found?

....

Figure 5.5 Button

What are the seven parts of the interface on the QGIS Project screen?

1. ....

2. ....

3. ....

4. ....

5. ....

6. ....

7. ....

5.2   Rasters and Vectors

There are two kinds of data you can use in QGIS: Raster and Vector. We are going to have a look at the differences between them.

Use the browser panel to navigate to where you have saved the supporting materials and find:

'Collymoon_SSSI_boundary.shp'
'NS59NE.TIF'

Figure 5.6 Navigate to layers in browser panel

The browser panel is similar to using Windows File Explorer or Finder, except that to get into a folder directory you left-click on the arrow to the left of the disk drive and a dropdown will appear, you will likely need to go through several folder dropdowns to reach your data.

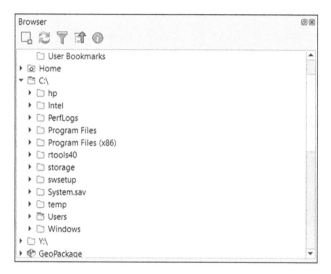

Figure 5.7 Navigate to layers in browser panel

Windows defaults to saving your data within C:\Users*Your username*\ Documents.

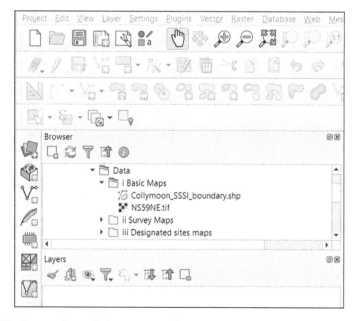

Figure 5.8 Navigate to layers in browser panel

Click and drag these into the layers panel below.

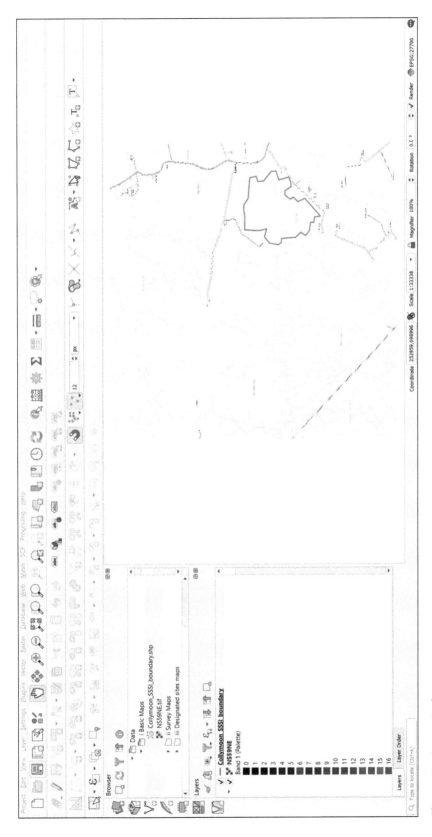

Figure 5.9 Layers in the map

The 'Collymoon_SSSI_boundary' is a line. This is a vector layer signified by the line icon (Fig 5.10) next to it.

The 'NS59NE' is a map image. The chequered symbol next to 'NS59NE' indicates that it is a raster layer.

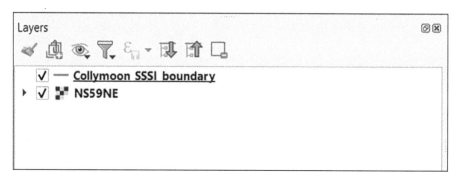

Figure 5.10 Layers in the layers panel

Note: If the boundary line is not showing up on the map but it is in the layers panel, it is either underneath the map or the line is too faint for the scale. You can change the position of layers in the layers panel by clicking and dragging them above or below each other. (We will change how vector layers look, or symbology of layers later on.)

Use the Save button or in the top menu Project > Save As in the top menu to save this set-up with the layers added. You have now created a QGIS Project containing two layers: one vector, one raster.

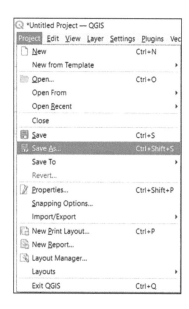

Figure 5.11 Save project

There are three types of Vector data: points, lines and polygons. This vector layer is a line and it indicates the site boundary. If we **right-click** on the layer in the layers panel a menu comes up. At the top of the menu is '**Zoom to layer**'; **left-click** on this to automatically zoom in on the boundary. **Right-click** on the layer to bring up the vector layer menu again. In the middle of this menu is 'Open Attribute table' – this can hold lots of information about the shape or shapes in the layer. A vector layer is editable in QGIS and it is the Attribute table that holds the information we put into it.

Figure 5.12 Vector layer menu

This raster layer is an Ordnance Survey map image or basemap. If we **right-click** on the layer a menu comes up. At the top of the menu is '**Zoom to layer**', **left-click** on this to automatically zoom out to the map extent. Notice no '**Open Attribute table**' option comes up. This is because there is no shape or location information about this raster layer; only vector layers have Attribute tables. Raster data is not editable in QGIS.

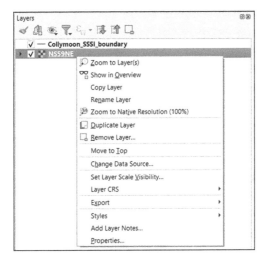

Figure 5.13 Raster layer menu

What is a QGIS Project?

A QGIS Project is a file that can be opened by QGIS Desktop. It displays layers.

What is a layer?

A layer is a group of files that together hold the coordinates or location information and data or attributes of the image you see on the screen.

Write down three things you notice about each layer:

Hint: see text above

Vector:

1. ....
2. ....
3. ....

Raster:

1. ....
2. ....
3. ....

For answers please go to Chapter 12: Answers to Exercises

5.3 Shapefiles and GeoTIFFs

Now we are going to have a look at where the files that make up the vector layer come from. For that we need to go to the folder containing the supplementary materials on your computer. Do this by minimizing QGIS and using Windows File Explorer or the Finder app on Mac to navigate to the folder.

Open the unzipped folder QGIS for Ecologists' and then the subfolder '5. Basic maps'.

In here there should also be a number of files of different types called '**Collymoon_ SSSI_boundary**'. These are the files that make up the '**Collymoon_SSSI_boundary**' layer we are viewing in the Project. We can open vector data in QGIS and edit it (more on this later).

Together the vector files are collectively known as a Shapefile. These are points, lines or polygons that display in the right place on the planet when opened in QGIS. The component files contain information about the layer and where in the world it is located. These files are vector data files that together comprise a shapefile and they are all necessary for the layer to work.

What is a Shapefile?

A shapefile is a group of files that comprise a vector layer. These files contain information on a shape and its location. Shapefiles can be viewed and edited in QGIS.

Rasters can be composed of one or more files. We can view the raster image in QGIS, but we cannot edit it. Together the raster files are collectively known as a GeoTIFF. This is an image in the right place on the planet when opened in QGIS, known as a georeferenced image file (more on georeferencing images later).

What is a GeoTIFF?

A GeoTIFF is an image with geographical information. GeoTIFFs can be viewed in QGIS, but attributes cannot be added (more on this later).

In your own words, define Vector and Raster data:

A Vector is:

....

....

....

A Raster is:

....

....

....

For answers please go to Chapter 12: Answers to Exercises

Congratulations!

You now know the differences between Raster and Vector data!

5.4 Map navigation

In order to let us move around the map there are a number of tools at our disposal. These are found within the buttons in the top toolbars. You will have noticed above that when you hover over a button it gives you an explanation of what it does.

Left-click to use each of the tools shown below and describe what the following buttons do in your own words, the first one is filled in for you:

Pan button – lets you move around the map.

Figure 5.14 Map navigation buttons exercise

5.5 Coordinate Reference Systems

In the bottom info bar in the right-hand corner of the screen you will notice letters with numbers after it. This is the Coordinate Reference (CRS) of the Project. This should have defaulted to ESPG:27700 which is the CRS of the shapefile: OSGB 36 / British National Grid ESPG:27700.

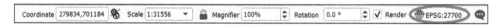

Figure 5.15 Coordinate Reference System

What is a Coordinate Reference System (CRS)?

A CRS is the way QGIS knows how to display the vector and raster data. Vector and raster layers are created with a CRS, but can be displayed in different CRS. For UK-based ecologists needing measurements in metres use British National Grid (for UK excluding Northern Ireland and The Republic of Ireland where Irish Grid also in metres, TM 65 / Irish Grid EPSG 29902, should be used). The other CRS you are likely to use is WGS 84, a world CRS measured in degrees around the globe. This is the CRS used by Google Earth and the default of GPS. Measuring tools and area calculations will not be in metres in this CRS (more on this later).

Click on **Coordinate Reference System** in the bottom right-hand corner. The Coordinate Reference System window should appear.

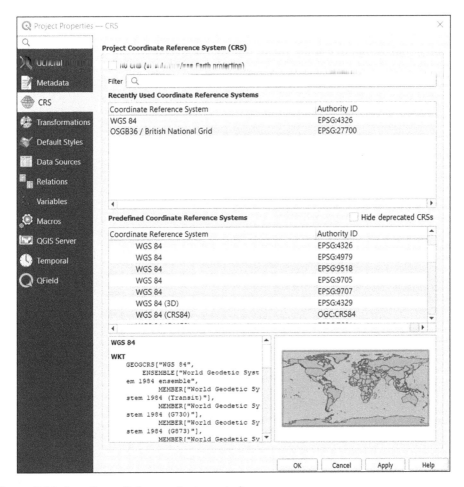

Figure 5.16 Coordinate Reference System window

In the search bar type '**3857**'.

Click on '**WGS 84 / Pseudo-Mercator EPSG: 3857**' a pop-up window asking for you to Select Transformations will appear. The default at the top should be already highlighted. **Scroll** across to the right to check this is the one with the highest accuracy i.e. accurate to 1m or less. Ensure the '**Make default**' box is ticked and **click 'OK'**.

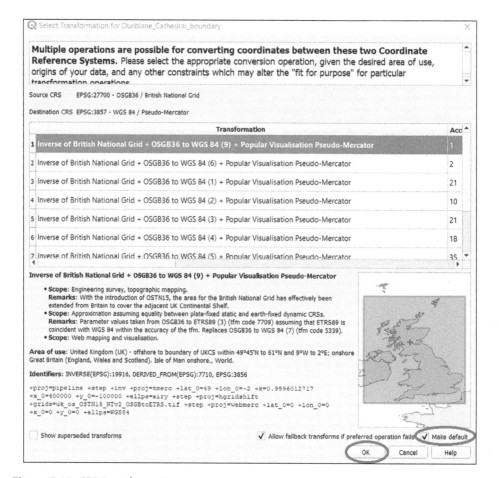

Figure 5.17 CRS Transformations

What are transformations?

Transformations are calculations that QGIS carries out in the background in order to help you convert between different units of coordinate reference systems and projections, e.g. between WGS84 in decimal degrees and OSGB36 in metres. Transformations differ in terms of accuracy: the lower the accuracy the higher the number quoted to the right of the transformation name in the table in the Select Transformations pop-up window. When quoting measurements in QGIS where a transformation has been applied, you need to bear in mind the accuracy of the transformation – for example, only quoting to the nearest metre if the accuracy is just to 1m.

Back in the CRS window **click 'OK'** to apply to the map and exit the Coordinate Reference System window.

What do you notice about the basemap?

....

Why has this happened?

....

Why did we need to select a transformation?

....

To change it back **click** on the **Coordinate Reference System** in the bottom right-hand corner. In the search bar type **"British" and press enter.**

Click on 'OSGB36 / British National Grid EPSG 27700' and 'OK' to apply to the Map and exit the Coordinate Reference System window.

5.6 Mapmaking basics

Now we understand the vector and raster data and have our site boundary and basemap layers, we are going to make a basic map.

Use the tools we have learned above to zoom in to the boundary leaving some area of basemap around the boundary for context.

Right-click on **'Collymoon_SSSI_boundary'** and **left-click 'Zoom to layer'.**

What has happened to the basemap image and why?

....

....

Zoom out using the 'Pan' tool until the basemap is in focus, keeping the boundary in the middle of the map screen.

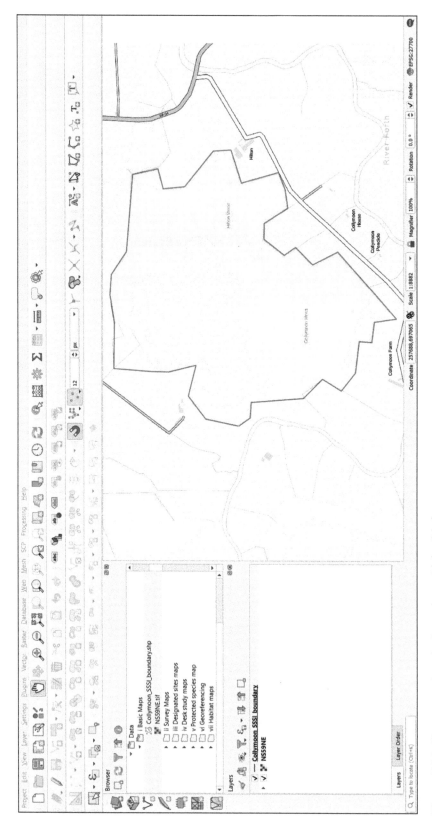

Figure 5.18 Map centred on the boundary with legible basemap

Click on the 'New Print Layout' button in the top toolbar.

Figure 5.19 New print layout button

Give your layout the title "Collymoon SSSI map".

You see a new window interface with a blank page. Maximize the window using the square button at the very top right of the window.

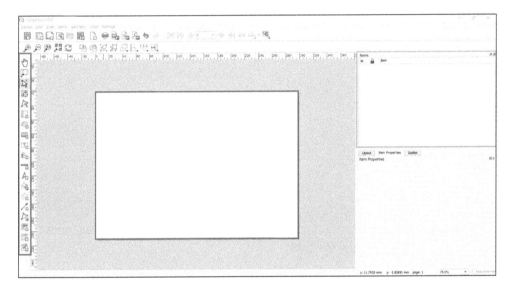

Figure 5.20 Blank layout screen

You will notice there are familiar map navigation buttons in the top toolbar.

Use the 'Zoom Full' button to make the white area or canvas larger.

Label the four areas of the layout view highlighted in Figure 5.20 above:

1. ....

2. ....

3. ....

4. ....

On the left-hand side is a toolbar. As you use these buttons in the following exercise, write down what each tool does next to the image:

Figure 5.21 Layout left toolbar

Hover over these working your way down the toolbar until you come to '**Add new map**'.

Click on '**Add new map**'.

Click at the top left-hand corner of the blank page and **hold** the left mouse button down as you **drag** a box across the page and **lift** your finger to finish. Resize if necessary to ensure box fits within canvas.

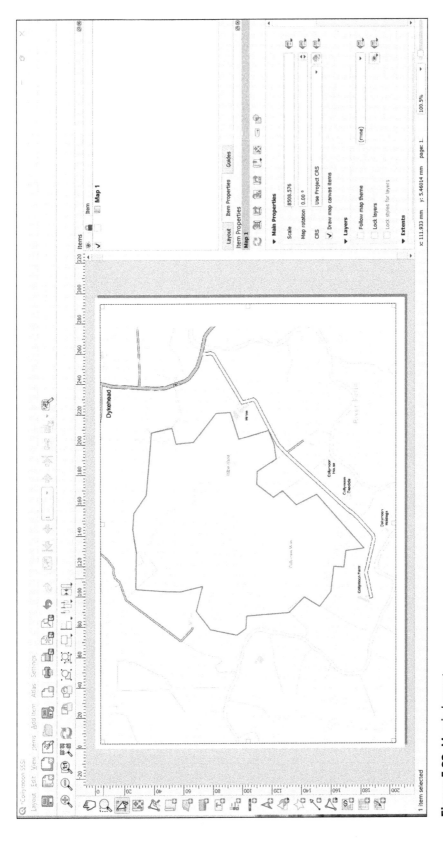

Figure 5.22 Map in Layout

You now have a map, let's get it ready to print.

Use the **select button** in the left toolbar and **click** on the map to select it. A list of Item properties should now appear in the right-hand panel.

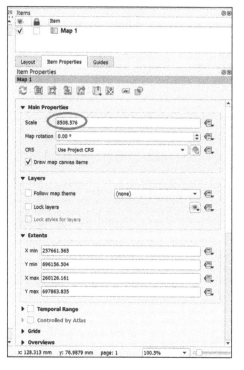

Figure 5.23 Item properties right panel with scale circled

What is scale?

Scale is how much detail is present on the map and how close we need to zoom into it to make the text legible. A small-scale map has little detail and a large-scale map has much more detail. In this case scale is measured in metric units with 1 centimetre = 10,000 metres.

Type an appropriate scale for the map, in this case: 10000.

Go back to the left-hand side toolbar and find the 'Add new scalebar' button.

Circle and label this on the previous page.

Click on 'Add new scalebar' then **click** draw a box for it in the bottom left corner of the map.

The scalebar can be adjusted using the 'Item properties' tab on the right-hand side. If this does not appear use the 'Select icon' to select the scale bar you have added to the map. Scroll down in the 'Item properties' menu and click on the box beside 'Background' to add a fill to the scalebar to map it more visible.

Ensure you have the Map selected when you add in the scalebar, or select 'Map 1' in the scale bar 'Item properties' dropdown.

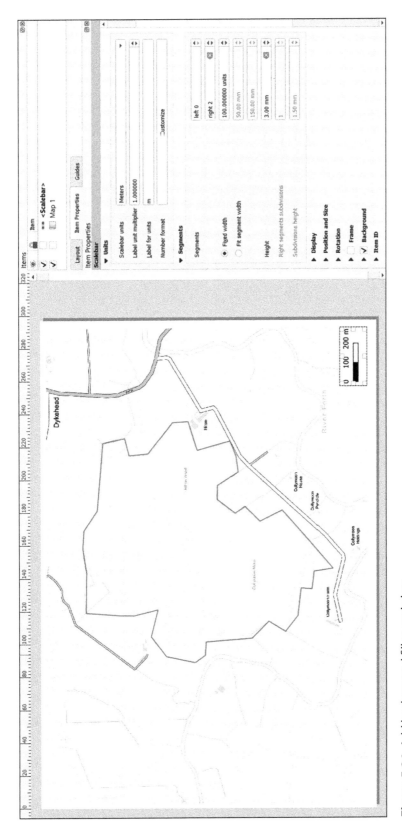

Figure 5.24 Add background fill to scale bar

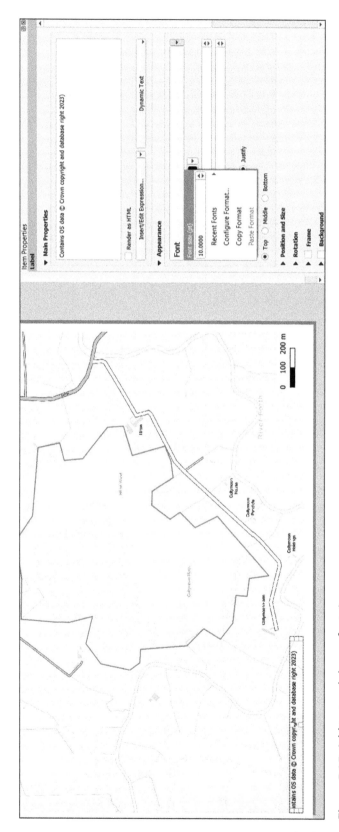

Figure 5.25 Add text and change font size

We are going to add a copyright statement to the map.

Copyright

Whenever you print a map it is essential to add a copyright statement citing all the sources of data you have used to create it. This is a legal requirement to comply with the licence conditions (more on copyright later).

Use the 'Add new label' button and draw the position of the text box at the bottom left corner of the map.

Add copyright text to the Label > Main Properties box in the right panel using the current year e.g. 2023: "Contains OS data © Crown copyright and database right (year)". This is for the raster basemap. We also need to add a copyright statement for the vector dataset of the site boundary this is: "© SNH, Contains Ordnance Survey data © Crown copyright and database right (2022)".

Change the text size to 7 and add a background fill to the box.

Resize the scalebar and copyright text boxes by hovering over the boxes and using the arrows to drag the edges of the boxes downwards.

Circle and label the buttons below:

Figure 5.26 Layout top toolbar

Go back to your answers for the labelling of the four areas of the layout view and add a description of what each of the areas is, now that you have used them.

Now we can export our map to print later.

Hover over the top toolbar until you come to 'Export as image' and choose where on your system and in which format (JPEG or PNG are good for putting into reports) you wish to save the file. Alternatively, you can 'Export as pdf' for a stand-alone image.

A box will pop-up with export resolution options.

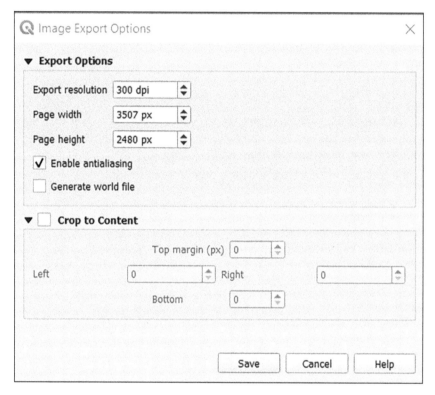

Figure 5.27 Layout screen export resolutions window

300dpi is fine for exports up to A4 size; increase the dpi for larger images or decrease for smaller.

Great! You have now made a basic map!

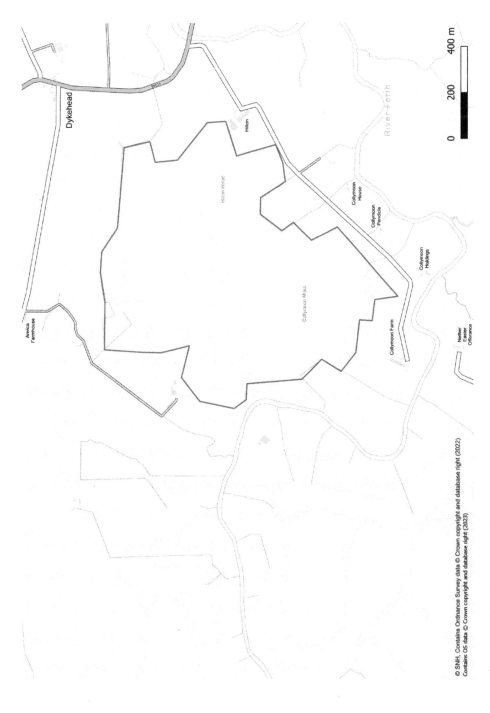

© SNH, Contains Ordnance Survey data © Crown copyright and database right (2022)
Contains OS data © Crown copyright and database right (2023)

Figure 5.28 First map export

5.7   My Instructions for... exporting a basic map using QGIS

List the six steps you used to produce your basic map:

1.   ....

2.   ....

3.   ....

4.   ....

5.   ....

6.   ....

5.8   Downloading basemaps from OS Open Data

The raster basemap provided for the previous exercise 'NS59NE.tif' is an Ordnance Survey map at 1:25,000 scale downloaded from the Ordnance Survey Open Data website. The Ordnance Survey Open Data website is useful for free images on a small-scale for the UK (excluding Northern Ireland).

From the website you can download raster basemaps at different scales as well as vector datasets. If you would like a higher zoom level for your basemap and therefore a higher quality final image, consider purchasing the larger-scale 1:10,000 basemaps or the Mastermap vector data license (if you work in the public sector you may have access to these already).

Go to the OS Open Data downloads page and download data for your local area or the area you are working in by Ordnance Survey grid square. At time of writing the website is: https://osdatahub.os.uk/downloads/open

To download maps at the same scale as the Collymoon SSSI example 1:25,000 scale on the websites these are listed as '**OS Open Map – Local**'.

You can either download the whole of the GB dataset or use the 'Set Custom Area' select one or more map tiles and **click** 'Save selection'.

To download the raster images, make sure to select '**GeoTIFF Full Colour**' from the dropdown menu.

Click to download and a pop-up window will appear. **Select** 'Save file' and 'OK'.

Ordnance Survey National Grid/British National Grid

Ordnance Survey National Grid splits the UK into 55 squares of 100 × 100km described by two letters. For example, in the name of the Collymoon SSSI basemap we used 'NS59NE.tif', the first two letters indicate it is within square 'NS'.

5.9   Making a map of your site

Unzip the folder you have downloaded. **Open** a new QGIS Project. **Set the CRS** to OSGB 36/British National Grid ESPG:27700. Add the basemaps by navigating to them using the Browser panel.

You can select all of the raster map layers using the 'Shift' key on your keyboard and drag them in all at once.

Once they have loaded, **right-click** on one of the raster layers and click '**Zoom to layer**'. Use the '**Pan' and 'Zoom' tools** to find a local site of interest to you, or a site you are currently working on.

Remove the layers related to **Collymoon SSSI** by **right-clicking** on them and **left-clicking** on 'Remove layer'. **Save as** a new Project titled with the site name.

Use the steps in '**5.7: My Instructions for... exporting a basic map using QGIS**' to create a map of your site without a site boundary. (We will look at creating vector data later in the book, so for now just export the raster map.)

Well done! You have now created a basic map for your own site!

5.10   My Instructions for... downloading and loading basemaps

What are the seven steps for downloading and loading basemaps into QGIS?

1. ....

2. ....

3. ....

4. ....

5. ....

6. ....

7. ....

6. Survey maps

In order to survey a site you first need to have a map of that site. This needs to be at a sensible scale to navigate by and draw on in the field. By the end of this chapter you will learn:

- How to use online basemaps.
- How to produce a survey map for field use.
- How to produce an aerial imagery map for field use.

6.1   Connecting to online basemaps

Now we understand the QGIS interface and the difference between raster and vector data we are going to make a survey map. For our survey map we need a basemap at a larger scale than our Ordnance Survey basemap.

> The online basemaps or Web Map Services detailed here are free and open source for use commercially in QGIS. However, many cannot be used commercially – for instance Google and Bing. Always check the license conditions on the website before producing a map for publication, e.g. in a report.

To do this we are going to use an XYZ Layer connection. You will need to be connected to the internet to use XYZ Layer for this exercise.

> *What are XYZ Layers?*
>
> XYZ Layers are basemaps and other data provided via the internet with dynamic scaling so you don't need to download separate basemap tiles and can merrily zoom in and out at will.

Open your "Collymoon SSSI" QGIS Project from the previous chapter.

Find '**XYZ tiles**' in the **Browser panel** and use the arrow on the left to open the dropdown.

Figure 6.1 Add XYZ connection from Browser panel

Click on and drag 'OpenStreetMap' into the Layers panel or **double-click** to automatically add.

You may get the **Select Transformations** window appear again, if so **follow** the instructions in '5.5 Coordinate Reference Systems'.

Drag the layers in the Layer panel so the basemaps are under the vector layer.

Untick 'NS59NE' or drag to the bottom of the layers.

Zoom in and out of the map using the map navigation tools in the top toolbar.

What are the differences between using a downloaded Raster tile (i.e. the Ordinance Survey map) and XYZ tiles (i.e. the OpenStreetMap)?

....

....

6.2   Survey mapmaking

We are now going to map our data, as we did in the previous chapter. Look back your notes in '5.7 My instructions for making a map using QGIS'.

What kind of data is OpenStreetMap and how can you tell?

....

Right-click on 'Collymoon_SSSI_boundary' layer and **left-click** on 'Zoom to layer'

1. **Open** 'New Print Layout' button in the top toolbar and title **"Collymoon SSSI OSM map".**
2. **Zoom in** so the canvas fills as much of the page as possible. **Click** on 'Add new map', then click at the top left-hand corner of the canvas and drag a box across the page and then double click. Ensure map box is fully within the canvas.

3. **Adjust scale: Click** on the map and in the right-hand panel under 'Item properties' type an appropriate scale for the map, in this case: "8000".
4. **Delete old scalebar and add new scalebar**. Adjust using the 'Item properties' tab on the right-hand side. Ensure you have the Map selected when you add in the scalebar or select 'Map 1' in the scale bar 'Item properties' dropdown.
5. **Add copyright information:** Use the '**Add new label**' **button** and draw the position of the text box with the left mouse button on the canvas. Format the text under the Item properties tab. **Add text**: "© OpenStreetMap contributors". Note: this differs from previous chapter copyright statement because the basemap comes from a different source.
6. **New Step: 'Add a north arrow'.** For the survey map we are going to add a north arrow, so we know which way is north for navigation in the field. Left-click on the '**Add north arrow**' **button** and draw a box for the position you want for the north arrow. You should now see a north arrow where you have drawn the box.

You can choose different arrows from the 'Item properties' tab in the right panel.

As another aid to surveying we are going to add grid-lines onto our field map. This is also useful if we draw on the map in the field and then want to georeference our image latter to digitize from (see Chapter 10: Georeferencing maps).

7. **New Step: 'Add grid-lines'. Click** on the map and in the right-hand panel under 'Item properties', scroll down to 'Grids' and use the **green plus** button to add a new grid.

Figure 6.2 Add grid

Then **click** the 'Modify grid...' button. In the next tab, change the X and Y interval from **0 to 100**. This gives you grid-lines at 100m separation.

Figure 6.3 Change grid interval

Scroll down and **tick** the '**Draw coordinates**' box. Change the left and right formatting to '**Vertical ascending**'. Change '**Coordinate precision**' to 0. Reduce the font size. Adjust the size of your map on the page to ensure the grid coordinates will fit on the page.

Figure 6.4 Add coordinates

8. **'Export as image'** and choose where on your system and in which format.

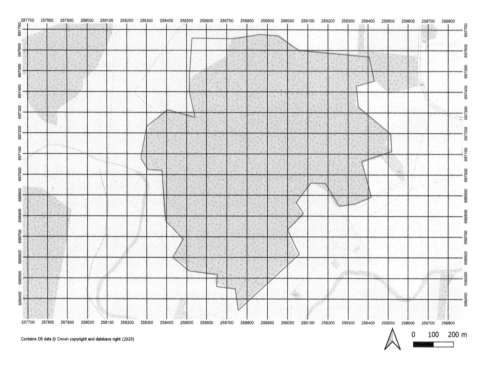

Figure 6.5 OSM map export with grid

Brilliant! You have now made a survey map!

Open the Project you created in '5.9 Making a map of your site'. Repeat the process of connecting to OpenStreetMap. Zoom to a desired scale. Follow the steps in '6.2 Survey mapmaking'.

Good work! You have now made a survey map of your site!

6.3   My Instructions for... connecting to online basemaps

Describe in your own words how to connect to online basemaps:

....

....

6.4   Connecting to online aerial imagery

Instead of a cartographic basemap, we may prefer to use an aerial image for the basemap, to make finding and mapping habitats easier in the field.

Open a new QGIS Project and **Save as** 'Collymoon SSSI Aerial'.

Navigate to the Collymoon SSSI boundary in the Browser panel and **drag** into the Layers panel.

We are going to add in a free aerial Web Map Service from ArcGIS. Unlike with the OpenStreetMap connection that comes with QGIS, we need to first add a new connection before we can add in the imagery.

Web Map Services are hosted on websites whose addresses often change, so it is always best to check these and the accompanying licenses to ensure you are working with the most up-to-date connections and within copyright laws by reading the licence materials for the layer you wish to use.

Open a web browser and search for World Imagery. At time of writing the website is: https://www.arcgis.com/home/item.html?id=10df2279f9684e4a9f6a7f08febac2a9

On this page you will find the description, citation and url of the host service.

Copy and paste the citation and url of these into a document for later use.

Back in QGIS:

Click on '**Open Data Source Manager**' in the left toolbar.

Scroll down and Click the '**ArcGIS REST Server**' tab from the left side of the window.

Click 'New'

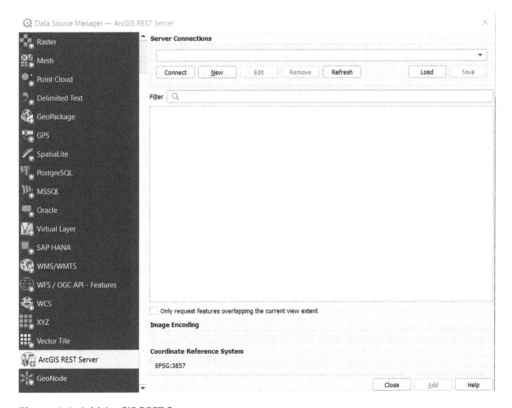

Figure 6.6 Add ArcGIS REST Server

Give a name for the connection, in this case 'World Imagery'.

In the URL box paste in the URL you copied from the website, at time of writing this is: https://services.arcgisonline.com/ArcGIS/rest/services/World_Imagery/MapServer

Click 'OK'

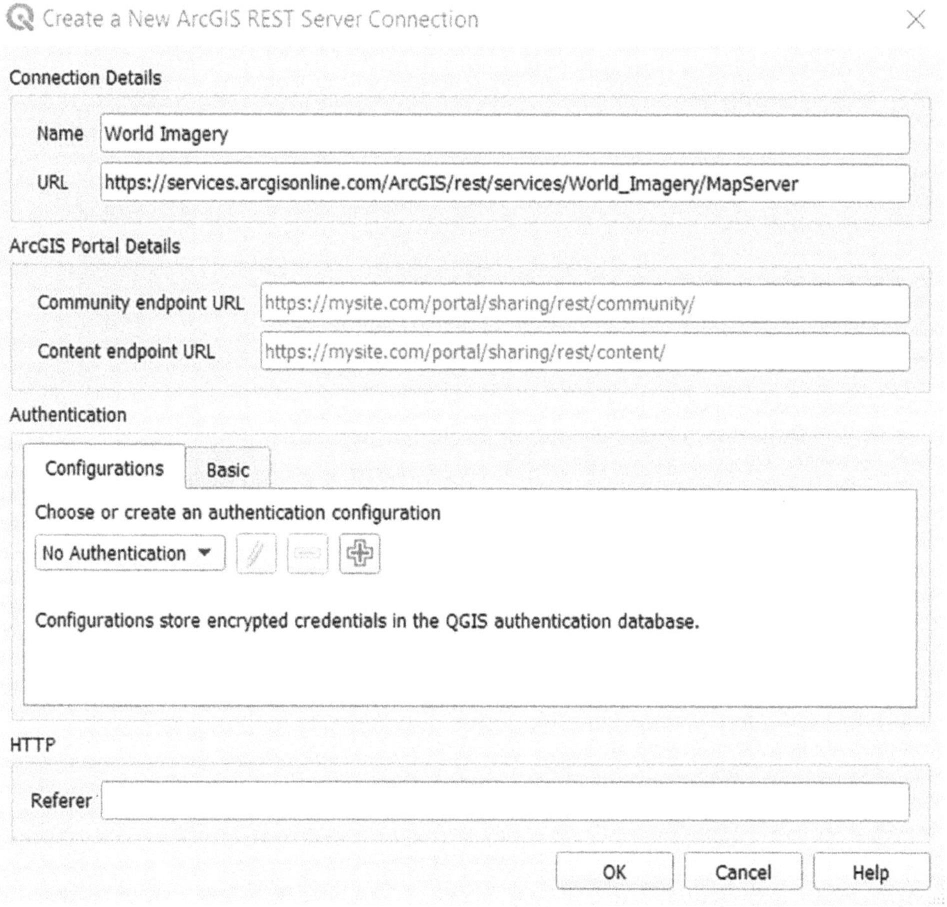

Figure 6.7 Create a New ArcGIS REST Server Connection window

Back in the **Data Source Manager** window **click 'Connect'**

Select 'World Imagery' and **click** on the Raster layer '**1.2m Resolution Metadata**' then 'Add'.

Note: It is not recommended to add in all the layers at once as the size of these tends to slow down or crash QGIS. Try the lowest resolution one, as detailed above.

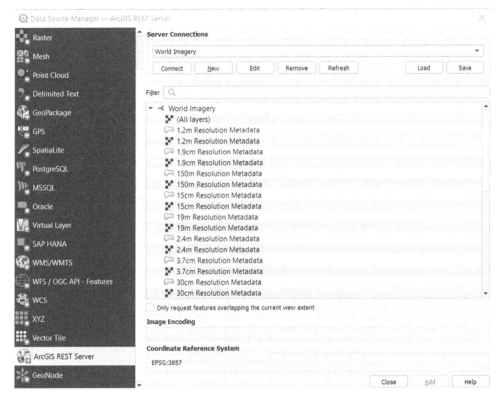

Figure 6.8 Connect to world imagery and add to map

Wait for imagery layer to load.
Drag the boundary above the aerial layer.

6.5   Aerial Imagery mapmaking

Follow the steps from the previous exercise '6.2 Survey mapmaking' to make an aerial survey map for use in the field – **changing the copyright statement** to the **citation** you copied from the website. At time of writing this is:

"Source: Esri, Maxar, GeoEye, Earthstar Geographics, CNES/Airbus DS, USDA, USGS, AeroGRID, IGN, and the GIS User Community"

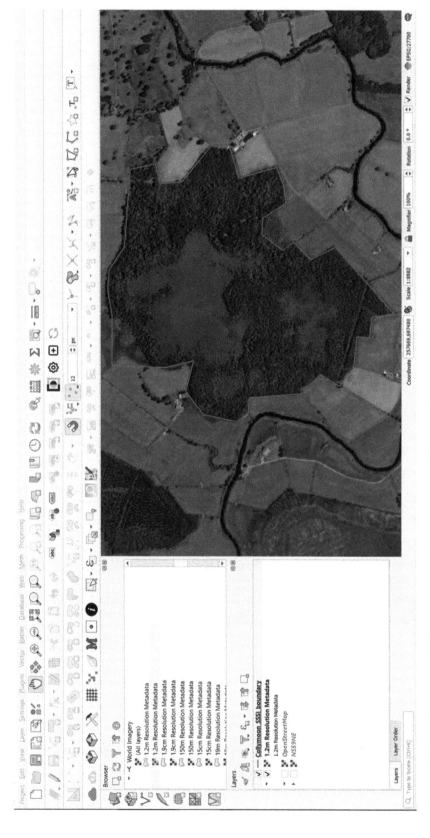

Figure 6.9 Aerial imagery in QGIS cannot be used as a basemap without purchase

Under the terms of the license this imagery can only be used for visualization and digitization and not used as a published basemap. It can be used in the field for data collection but not in report maps.

Very good! You have now made an Aerial imagery map!

Open the QGIS Project you created in '5.9 Making a map of your site'.

Now we have established the connection to the imagery it is now available in the Browser panel.

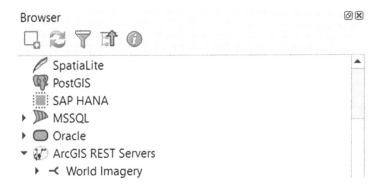

Figure 6.10 Add imagery with established connection

In the **Browser panel, Click** the arrow at the left side of 'ArcGIS REST Service' to show dropdown menu then Click the arrow at the left side of 'World Imagery' to show the dropdown submenu.

Double-click on or click and drag '**1.2m Resolution Metadata**' into layers panel to add to map.

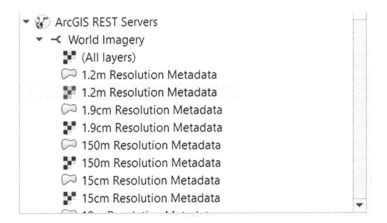

Figure 6.11 Add imagery with established connection submenu

Zoom to a desired scale. **Follow** the steps in '6.2 Survey mapmaking'.

Congratulations! You have now made an Aerial imagery map of your site!

6.6   My Instructions for... exporting a survey map using QGIS

List the eight steps you used to produce your survey map:

1.   ....

2.   ....

3.   ....

4.   ....

5.   ....

6.   ....

7.   ....

8.   ....

6.7   My Instructions for... basemap copyright

Why is it important to add a copyright statement to every map you produce?

....

What is the copyright statement for Ordnance Survey maps and data?

....

What is the copyright statement for OpenStreetMap?

....

What is the copyright statement for ArcGIS World Imagery?

....

....

Finding copyright statements

Right-click on the layer and click 'Layer Properties'. For how to display the copyright statement correctly, the full licences conditions including permitted data uses can be found on the copyright webpages of the data supplier. For downloaded data, licence information and copyright statements are usually found in the 'Read Me' documents within the same folder that the downloaded data comes in.

7. Designated sites map

In order to understand the significance of the site and any potential impact of works on that site, we need to check if there are any designated sites on or near our own site. You have downloaded a set of spatial data files (called a shapefile) of Sites of Special Scientific Interest (SSSIs). You now want to map this, get information on distances from your site to these imagined designated sites and the area of each of the SSSIs.

You will learn:

- How to set up a Project.
- How to perform basic analysis on vector data.
- How to produce a desk-study map from existing data.
- Where to download designated sites data.
- How to connect to web map/feature services.

7.1 Setting up a Project: Setting the Coordinate Reference System

Open a new QGIS project.

First, we need to set the Coordinate Reference System (CRS).

Why do we need to set the Coordinate Reference System?

By giving a Project a CRS we tell QGIS where on the planet to display vector and raster data we load into the Project. There are different CRS for the world scale and individual country scale. It is more accurate to use the national CRS for the country you are in rather than a world CRS. For mapping work in the UK (excluding the Ireland) the CRS is OSGB36 / British National Grid EPSG 27700. For mapping work in Northern Ireland and Ireland the CRS is TM 65 / Irish Grid EPSG 29902.

In the top toolbar **click 'Project'** then **'Properties'** from the menu then the **'CRS' tab** from the 'Project Properties' window.

Click on 'OSGB36 / British National Grid EPSG 27700' then 'OK'.

What is a coordinate reference system (CRS)?

....

....

....

....

Why is it important to set a CRS when setting up a project?

....

....

....

....

7.2   Setting up a Project: Importing vector and raster data

We have seen how to use the Browser panel to import data. Here is another method you may find more useful if, like myself, you have rather a lot of folders to navigate:

Into your Project **add** "Edinburgh_Castle_boundary.shp" and "SSSIs_SCOTLAND. shp" by using top toolbar **Layer** menu then **Add Layer then Add vector layer.** Or use the left shortcut bar **Open Data Source Manager** and select the **Vector** tab. (Either will take you to the same window then click on the three dots to navigate to the shapefile.)

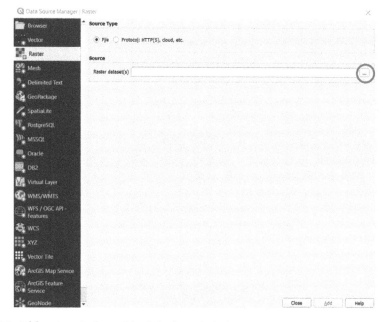

Figure 7.1 Add raster window with triple dots circled

Add in 'NT27SE' and 'NT27SW' by using the top toolbar **Layer** menu then **Add Layer** then **Add raster layer**, or the left shortcut bar **Data Source Manager** and select the **Raster** tab and click on the three dots to navigate to the GeoTIFF.

OR add in the OpenStreetMap basemap.

Reorder layers in the layer tab with basemap below vector layers.

Right-click a vector or raster layer and select '**Zoom to layer**' if you end up too zoomed out or pan off the screen.

Save project as "Edinburgh Castle SSSIs".

What is the difference between a shapefile and a project?

....

....

....

....

What is the difference between a map and a project?

....

....

....

....

7.3   My QGIS Instructions for... setting up a Project

What are the four steps for setting up a project?

1.   ....

2.   ....

3.   ....

4.   ....

7.4   Using QGIS tools

Use the identify features tool to find out the name of the SSSI located closest to the site boundary:

Find the '**Identify features**' tool by hovering over the top toolbar.

Use it by **selecting** the layer you want to identify in the layers panel, in this case 'SSSIs SCOTLAND' then clicking on '**Identify features**', then **click** on the SSSI shape closest to the castle boundary to identify it.

What is the SSSI name?

....

What do you notice when you use the identify features tool on the raster layer?

....

....

Find and use the '**Measure**' tool with '**Measure line**' selected to measure the minimum distance between the site boundary and the three closest SSSIs:

_____SSSI _____m from site boundary

_____SSSI _____m from site boundary

_____SSSI _____m from site boundary

Use '**Measure Area**' to measure the area of the three closest SSSIs:

_____SSSI _____ha

_____SSSI _____ha

_____SSSI _____ha

You can select points, lines or polygons in shapefiles using the '**Select**' tool.
Click on the SSSIs layer in the layers panel.
Click 'Select Features by Area or Single Click' and **click** on a SSSI on the map.
What happens?

....

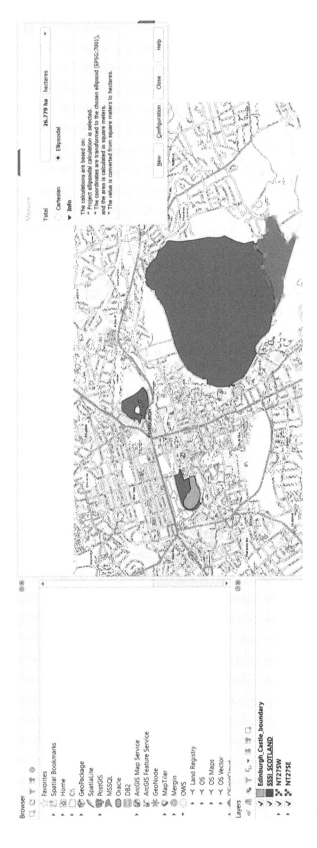

Figure 7.2 Measure area

Right-click on the layer for the SSSIs in the layers panel and **left-click** on 'Attribute table'.

You now see a table containing the SSSIs. To see only the SSSI you have selected in the map view, **click** on the filter dropdown menu and choose 'Show selected features'. You should now have a table with only your chosen polygon.

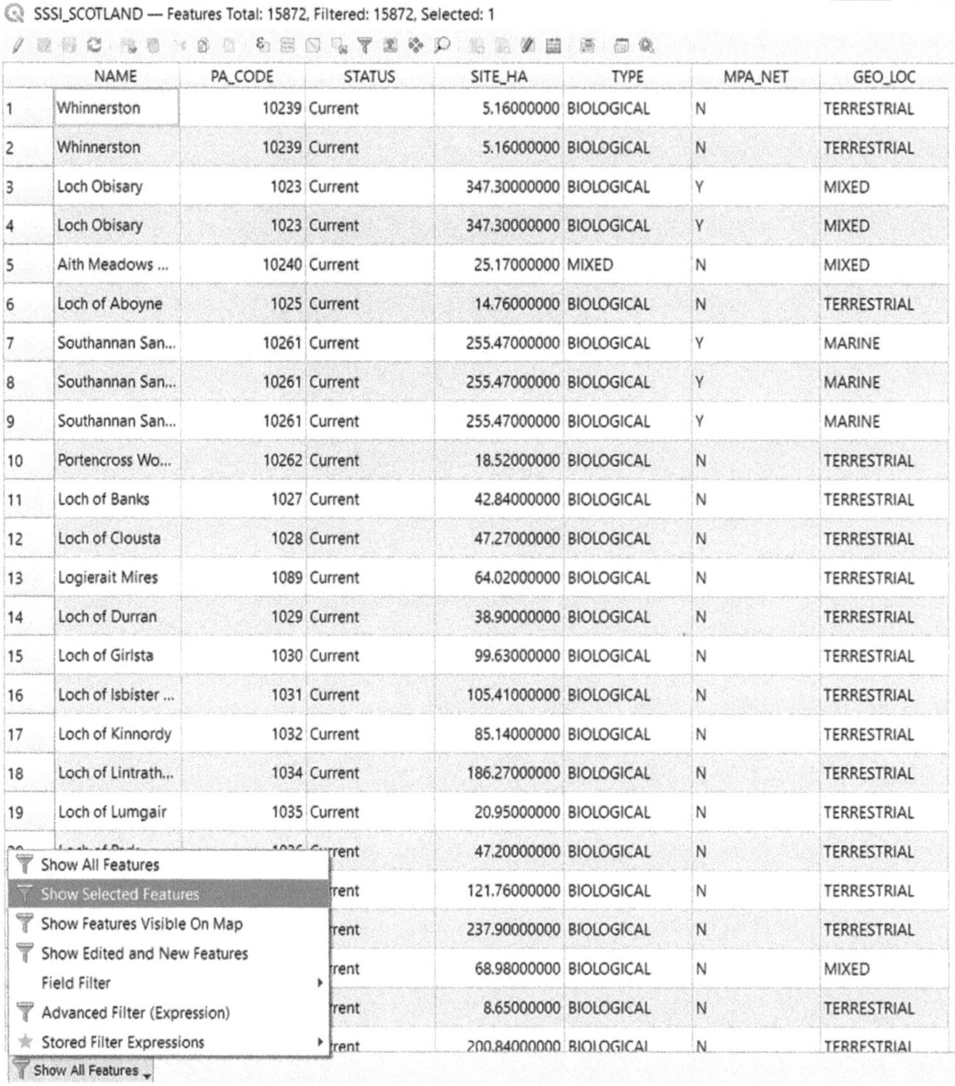

	NAME	PA_CODE	STATUS	SITE_HA	TYPE	MPA_NET	GEO_LOC
1	Whinnerston	10239	Current	5.16000000	BIOLOGICAL	N	TERRESTRIAL
2	Whinnerston	10239	Current	5.16000000	BIOLOGICAL	N	TERRESTRIAL
3	Loch Obisary	1023	Current	347.30000000	BIOLOGICAL	Y	MIXED
4	Loch Obisary	1023	Current	347.30000000	BIOLOGICAL	Y	MIXED
5	Aith Meadows ...	10240	Current	25.17000000	MIXED	N	MIXED
6	Loch of Aboyne	1025	Current	14.76000000	BIOLOGICAL	N	TERRESTRIAL
7	Southannan San...	10261	Current	255.47000000	BIOLOGICAL	Y	MARINE
8	Southannan San...	10261	Current	255.47000000	BIOLOGICAL	Y	MARINE
9	Southannan San...	10261	Current	255.47000000	BIOLOGICAL	Y	MARINE
10	Portencross Wo...	10262	Current	18.52000000	BIOLOGICAL	N	TERRESTRIAL
11	Loch of Banks	1027	Current	42.84000000	BIOLOGICAL	N	TERRESTRIAL
12	Loch of Clousta	1028	Current	47.27000000	BIOLOGICAL	N	TERRESTRIAL
13	Logierait Mires	1089	Current	64.02000000	BIOLOGICAL	N	TERRESTRIAL
14	Loch of Durran	1029	Current	38.90000000	BIOLOGICAL	N	TERRESTRIAL
15	Loch of Girlsta	1030	Current	99.63000000	BIOLOGICAL	N	TERRESTRIAL
16	Loch of Isbister ...	1031	Current	105.41000000	BIOLOGICAL	N	TERRESTRIAL
17	Loch of Kinnordy	1032	Current	85.14000000	BIOLOGICAL	N	TERRESTRIAL
18	Loch of Lintrath...	1034	Current	186.27000000	BIOLOGICAL	N	TERRESTRIAL
19	Loch of Lumgair	1035	Current	20.95000000	BIOLOGICAL	N	TERRESTRIAL
			rent	47.20000000	BIOLOGICAL	N	TERRESTRIAL
	Show All Features						
	Show Selected Features		rent	121.76000000	BIOLOGICAL	N	TERRESTRIAL
	Show Features Visible On Map		rent	237.90000000	BIOLOGICAL	N	TERRESTRIAL
	Show Edited and New Features		rent	68.98000000	BIOLOGICAL	N	MIXED
	Field Filter						
	Advanced Filter (Expression)		rent	8.65000000	BIOLOGICAL	N	TERRESTRIAL
	Stored Filter Expressions		rent	200.84000000	BIOLOGICAL	N	TERRESTRIAL
	Show All Features						

SSSI_SCOTLAND — Features Total: 15872, Filtered: 15872, Selected: 1

Figure 7.3 Show selected features in Attribute table

What information do you see in the table for your chosen SSSI?

....

....

....

Figure 7.4 Attribute table showing only selected feature

Right-click on the top toolbar and **click to tick** the box beside 'Annotation toolbar'; a new set of buttons will now appear in the top toolbar. Find and use the 'Text Annotation' tool to label the SSSIs by selecting the button and double-clicking on the screen where you would like your annotation to appear.

The annotations should be located within or adjacent to the SSSI you want to annotate.

Type the name of the SSSIs in the annotation boxes you create.

Figure 7.5 Text annotation

7.5 Adding an Attribute field

We are going to add a field to the attribute table of "Edinburgh_SSSIs" with the distances we have measured from the site boundary.

Right-click on 'SSSIs SCOTLAND' in the layers panel and select "Open Attribute table". In the top of the Attribute table **click** on '**Toggle editing**'.

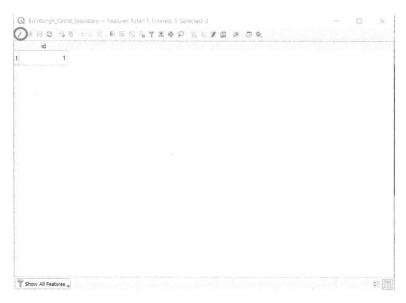

Figure 7.6 Toggle editing in Attribute table

Find the '**New field**' button at the top of the Attribute table and click.

We want to add a field with the title "**Distance**" of an integer type (whole number) of the length (in characters) of our measured distances from the site boundary.

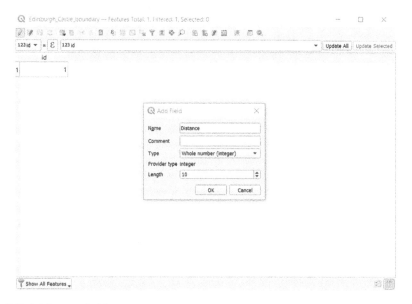

Figure 7.7 Add a new field

Fill in the details of the distances from the site boundary for each SSSI from your above records: **click** in each cell then **type** in the data.

Use the '**Identify features**' tool – we can now see the new information we added displayed in the attributes.

7.6 QGIS tools test

Describe what the following buttons do in your own words:

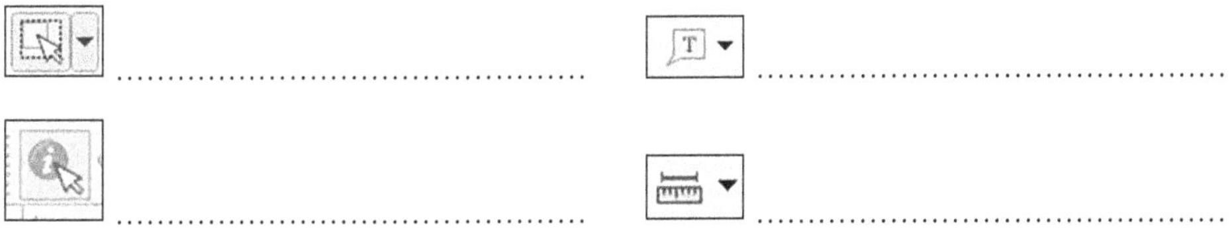

Figure 7.8 Tool buttons test

7.7 My QGIS Instructions for... measuring and recording distances

Describe how you measure distances between shapefiles and how to record them in the Attribute table:.

1. ....

2. ....

3. ....

4. ....

5. ....

6. ....

7.8 Styling lines and polygons

We are going to change the way the boundary and SSSIs appear in the map. QGIS calls this symbology.

To change the symbology (the way the shape is displayed) of a layer, **double-click** a layer in the layers panel (or **right-click** and select '**Properties**'). This opens the layer properties.

Choose the '**Symbology**' tab from the left-hand side of the pop-up box.

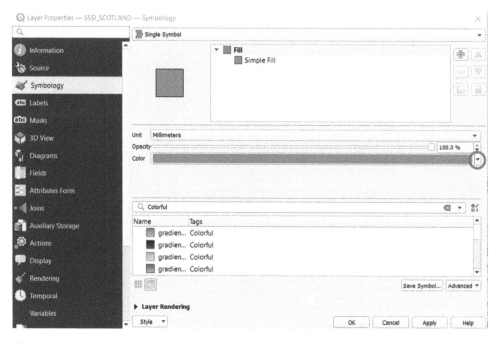

Figure 7.9 Symbology tab with colour dropdown circled

We can change the colour using the colour dropdown. This gives us a colour selector pop-up box where we can click to change the colour on the wheel and adjust the slider to reduce opacity/increase transparency. Have a go at changing the colour and transparency.

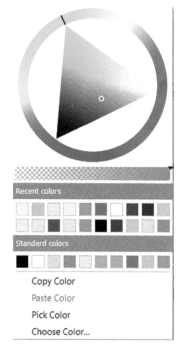

Figure 7.10 Colour selector

To get the full menu so we can change other things about the appearance of the layer:
Click on 'Simple fill' in the top box.
Choose 'Simple fill'.

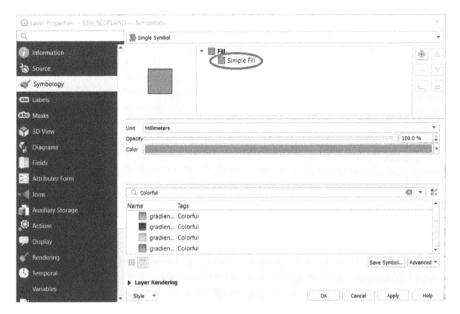

Figure 7.11 Symbology tab with simple fill circled

You will see we now have options to change the colour of the fill of the layer, but also its fill style and the stroke (outline) colour, width and style.

Figure 7.12 Symbology tab simple fill menu

Change the 'Edinburgh_Castle_boundary' to a red outline with no fill.

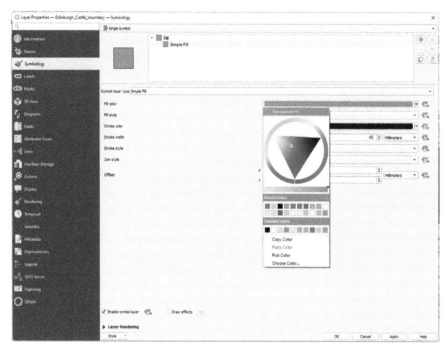

Figure 7.13 Symbology tab simple fill menu with transparent fill selected

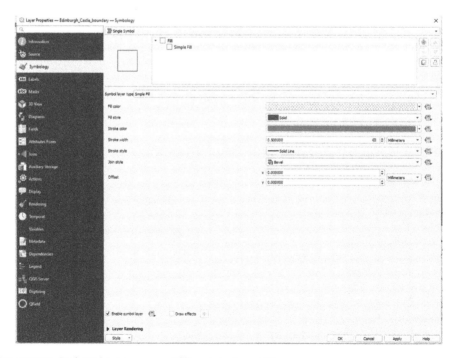

Figure 7.14 Red stroke, transparent fill and stroke width

Click 'Apply' and 'OK' to return to the map.

Repeat the process with 'SSSI_SCOTLAND', this time changing the Fill and Stroke to Black and the Fill style to 'FDiagonal'

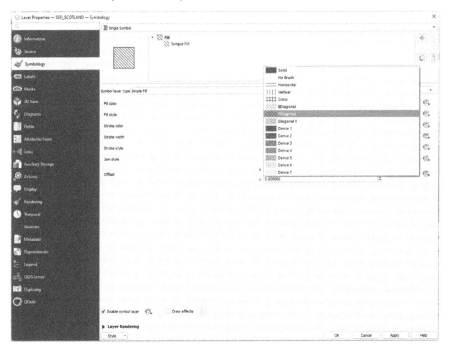

Figure 7.15 Symbology Simple fill, Fill style menu

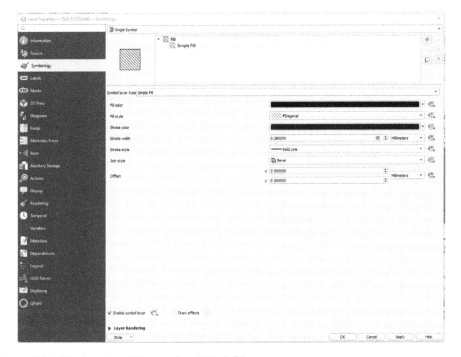

Figure 7.16 Black stroke, FDiagonal and Black fill

Click '**Apply**' **and** '**OK**' to return to the map.

Save project – This will save all changes.

Why is it useful to be able to change the appearance of vector layers in a map?

....

....

....

7.9   Labelling

Earlier in the chapter we looked at creating annotations, for this map we are instead going to use labels.

Go to the SSSI layer in the layer panel and **right-click** to open the layer menu and **left-click** to open the 'Properties' window. **Click** to select the 'Labels' tab. You will see we currently have labels turned off, use the dropdown to select '**Single labels**'.

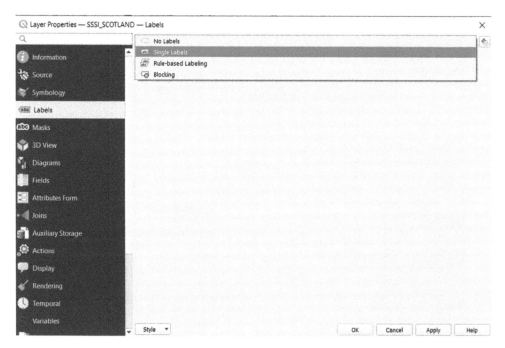

Figure 7.17 Labels tab with dropdown

We now have a menu of options for labelling the layer. This is where we can control a lot of features for labelling the layer. Select the 'Value' dropdown. This should have automatically selected 'Name'.

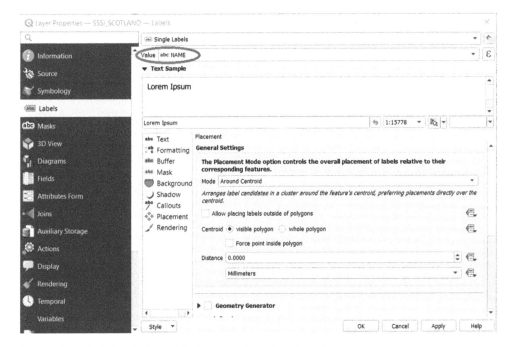

Figure 7.18 Labels tab single label menu with value circled

Open the 'Value' dropdown menu, what is this value and where does it come from?

....

....

Why is this useful?

....

Click 'Apply' and '**OK**' to return to the map, you should now see the SSSIs are labelled by name.

To change properties of individual labels including where individual labels plot, we can use the '**Label toolbar**' buttons.

Figure 7.19 Labels toolbar

Click on SSSI layer in the layer panel then Click on the 'Move label, callout or diagram' button.

Figure 7.20 Move label, callout or diagram

Hoover the mouse cursor over a label, what do you notice?

....

Click on a label.

A pop-up window will appear asking which field to store the new location of the point to. It does not matter which field you select from the menu as long as it contains a unique code to identify the row – in this case the 'Name' field is automatically selected and this is unique to the individual SSSI, so this is the option we want.

Click 'OK'

Figure 7.21 Label storage pop-up

The pop-up window will only appear the first time you want to move a label on the layer.

Click back on the **label** you want to move, then **click** where you want to move it to.

Repeat until the labels are where you would like them.

Compare the use of annotation and labelling, which do you prefer and why?

....

....

....

....

Annotations and labels

It is best to limit text to reduce clutter on the map. Short names or numbers work well. If further details are required these can be detailed in the legend, by inserting the attribute table into the map (using the button in the layout view) or in a separate page or appendix.

7.10   SSSI map

We are going to make a desk-study map using the "Edinburgh Castle SSSIs" project you have created.

Use the map navigation tools (pan, zoom, zoom to layer etc.) so you can see all the vector layers in your map view.

To make our Designated sites map we will follow the steps in '6.6: My Instructions for... exporting a survey map using QGIS' (see page 51), changing the following:

1. Open 'New Layout' and title "Edinburgh Castle Designations map".

5. Add copyright information: Add the copyright statement that matches the basemap you are using. We also need to include the copyright statement of the SSSIs, which is a Scottish Natural Heritage dataset: "© SNH, Contains Ordnance Survey data © Crown copyright and database right (2022)"

7. New step: Add a legend using the 'Add new legend' button. All legend entries will be added. Ensure legend is selected using the 'Select/Move item' button. Then we can modify as required under the 'Item properties' tab. First untick 'Auto update'.

Notice that the buttons under the 'Legend items' are no longer greyed out.

Select the basemap layers then the red minus button to remove these layers from the legend. The legend entries provided by these are not useful and very long.

Double-click on 'SSSI_SCOTLAND' in the 'Legend items'.

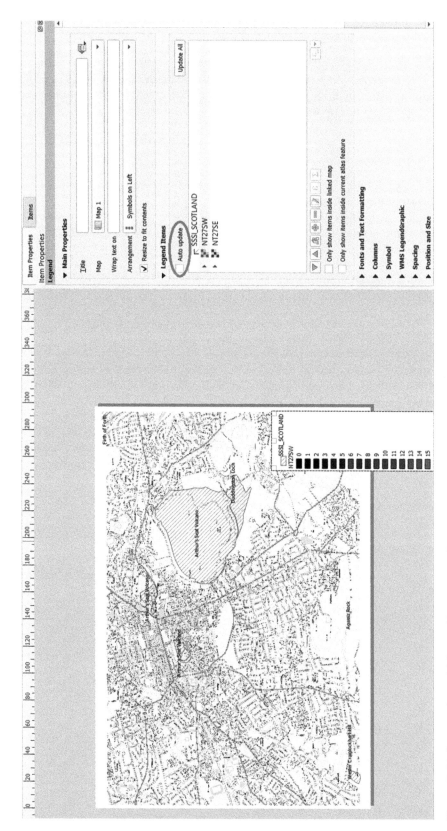

Figure 7.22 Layout view with legend Item properties

This opens a text edit box, **type** in "Sites of Special Scientific Interest (SSSIs)" as the new title for the layer.

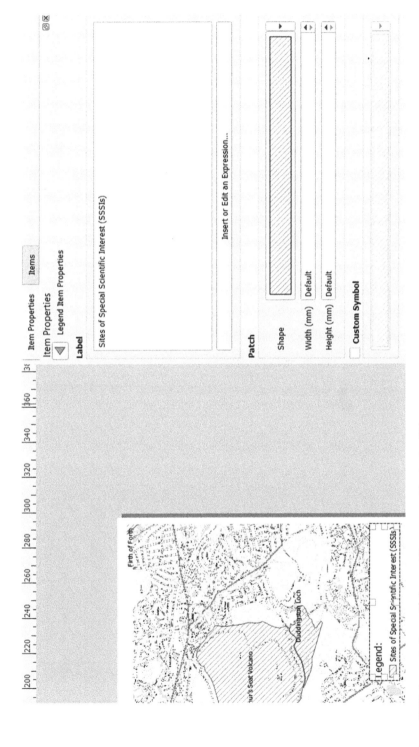

Figure 7.23 Layout view with legend Item properties text edit box

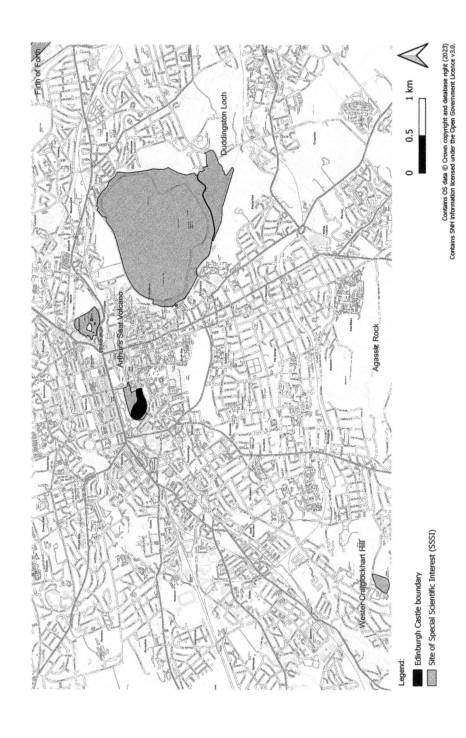

Figure 7.24 SSSI map export

Brillant! You have now made a SSSIs map!

7.11 My QGIS Instructions for... creating a map using styles and legend

Describe in your own words how to change the symbology of a vector file.

....

....

....

List the eight steps you used to produce your SSSI map:

1. ....

2. ....

3. ....

4. ....

5. ....

6. ....

7. ....

8. ....

Describe in your own words how to add and alter a legend in the Layout.

....

....

....

Accessing Designated Sites data

We are now going to repeat the above process after adding in Special Areas of Conversation (SACs) and Special Protection Areas (SPAs).

England and the Devolved Administrations have different organizations in charge of designations (as well as other environmental data).

As we have begun with an example in Scotland, section 7.12 below continues with the same example. Instructions for creating your own map in the other devolved administrations or in England, follow below in sections 7.14, 7.15, 7.16 respectively.

7.12   Scotland

Download layers:

Go to the NatureScot (formerly Scottish Natural Heritage) website and download the SAC and SPA shapefiles.

At time of writing the website is:

https://opendata.nature.scot/

Figure 7.25 Download Designated sites in Scotland

Search for 'SACs' and click through to the web map viewer page.

At time of writing the webpages for SAC and SPA download are:

https://opendata.nature.scot/datasets/snh::special-areas-of-conservation/

https://opendata.nature.scot/datasets/snh::special-protection-areas/

Under the '**Shapefile**' heading, **Click** on '**Download**'

Figure 7.26 Download SACs in Scotland

Unzip them.

Add to the map using your preferred method of adding vectors.

We now want to **change** the colours to be different from the SSSIs, as well as seeing what is going on underneath. One way of doing this is to alter the colour and direction of the hatching, e.g. changing the SPA layer Stroke to Blue and the Fill style to 'BDiagonal'.

Often, designations overlap each other – for instance, SSSIs legally underpin both SACs and SPAs.

Experiment with changing the symbology and zooming in and out on the map and your layers to see what looks best.

OR

As we did in '6.4 Connecting to Aerial Imagery' instead of downloading the shapefiles we can connect to via a url, but this time the host is not ESRI's ArcGIS Server, so we must set up the connection from a different tab in Data Source Manager.

At time of writing you find the url needed on to the WMS endpoints page on the NatureScot website: https://ogc.nature.scot/geoserver/protectedareas/ows

However, the endpoints change all the time, so below is the method for finding it that currently works, but the principles will remain similar:

Go to the NatureScot Open Data website and search for **'Protected Areas'**

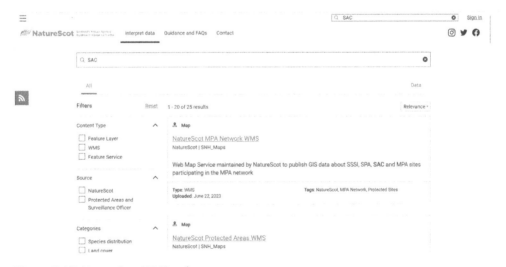

Figure 7.27 NatureScot WMS webpage

Open the NatureScot Protected Areas WMS web map viewer page and **click on the information button** to bring up a side menu.

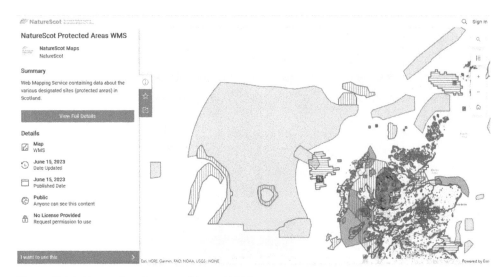

Figure 7.28 NatureScot Protected Areas WMS

At the bottom of the menu you will see '**I want to use this**'. **Click** on this, then on the next screen **click** '**View Data Source**'.

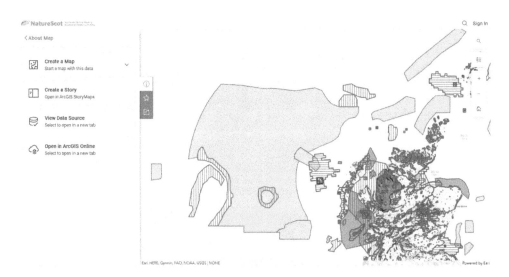

Figure 7.29 Select View Data Source

This will then bring you to the endpoints webpage.

Copy the web address from the web address bar.

Figure 7.30 Copy and paste web address

Back in QGIS, Click on '**Open Data Source Manager**' in the top toolbar.

Scroll down and left-click the '**WMS/WMTS**' tab from the left-side of the window.

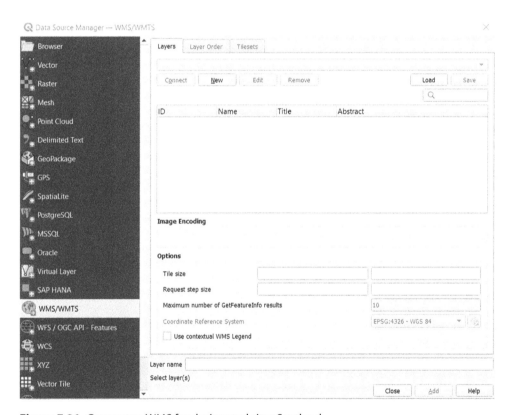

Figure 7.31 Connect to WMS for designated sites Scotland

Click 'New'.

Give a name for the connection, in this case "NatureScot".

In the URL box paste in the web address copied above, at time of writing this is: https://ogc.nature.scot/geoserver/protectedareas/ows

Click 'OK'.

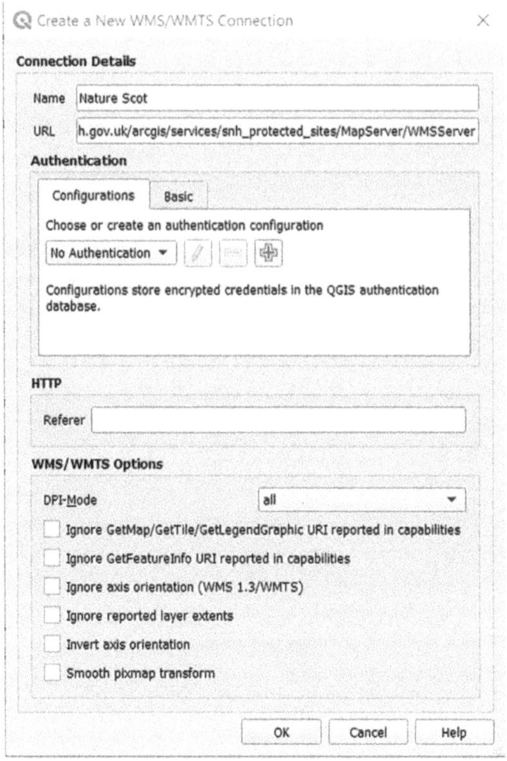

Figure 7.32 Create new WMS/WMTS connection for designated sites in Scotland

Back in the **Data Source Manager** window **click 'Connect'**

Select the layers then **click Add.**

In the Browser panel you should now see a huge list of layers when you open the NatureScot dropdown under WMS/WMTS. Scroll down in the layers panel to find and add into the map layer you want. However, it is often quicker to load and easier to search within the Data Source Manager connection to find the layers you want to load into the map.

Note: It is not recommended to add in all the layers at once, as the size of these tends to slow down or crash QGIS. Try one at a time.

Figure 7.33 Connect to SPAs layer

Wait for layers to load.

What do you notice about the layers?

....

....

To change how the layer looks, as before we can go into the Properties of the layer by right-clicking on the layer in the layers panel. Unlike in the downloaded vector layers, however, these are Web Map Service Raster datasets, so we cannot alter the symbology of the layers. But we can change the transparency of the layers in order to see what is beneath.

In the layer Properties select the '**Transparency**' tab.

Adjust the '**Global Opacity**' slider or type a percentage in the box.

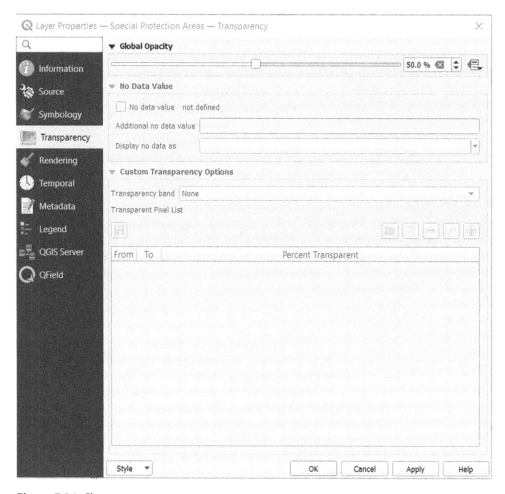

Figure 7.34 Change transparency

Click 'OK'.

What do you notice about the layer in the map?

....

Repeat for the other designated sites web map service layer.

> **What's the difference between a Web Map Service (WMS) and a Web Feature Service (WFS)?**
>
> Both provide online data you can connect to in QGIS. WMS are online raster services and WFS are online vector services. (See '5.2 Rasters and Vectors' for more about raster and vector data differences.)
>
> There does not currently appear to be a WFS available for Scotland.

Look in the bottom right corner of the QGIS window, what do you notice about the Coordinate Reference System (CRS)?

....

Why do you think this has happened?

....

Web Map/Feature Services use World Coordinate Reference Systems, for instance WGS 84/Pseudo-Mercator ESPG:3857. These have lower accuracy than Projected Coordinate Reference Systems such as British National Grid ESPG: 27700 that use metres and are therefore preferable for accurately measuring in metres and/or hectares.

7.13   Designated Sites Map

Follow the steps in '7.10 My QGIS Instructions for... creating a map using styles and legend' using either diagonal hatching and offsets or transparency of symbology to make a designated sites map changing the title of the Layout to "Edinburgh Castle Designated Sites map". The copyright statements will be the same as previously for the basemap and designation data.

Excellent! You have now made a designated sites map!

Open the Project you created in '5.9 Making a map of your site'. If your site is in Scotland, you can use the SACs and SPAs we have already downloaded. If relevant to your site you might want to also add RAMSAR, National Nature Reserves and Local Nature Reserves to the map.

Follow your instructions in '7.7: My QGIS Instructions for... measuring and recording distances' to measure the approximate distance from your site to the nearest SSSI, SAC, SPA.

Follow the steps in '7.10 My QGIS Instructions for... creating a map using styles and legend'.

Look up the copyright statement on the website where you downloaded the data from and add it to the map.

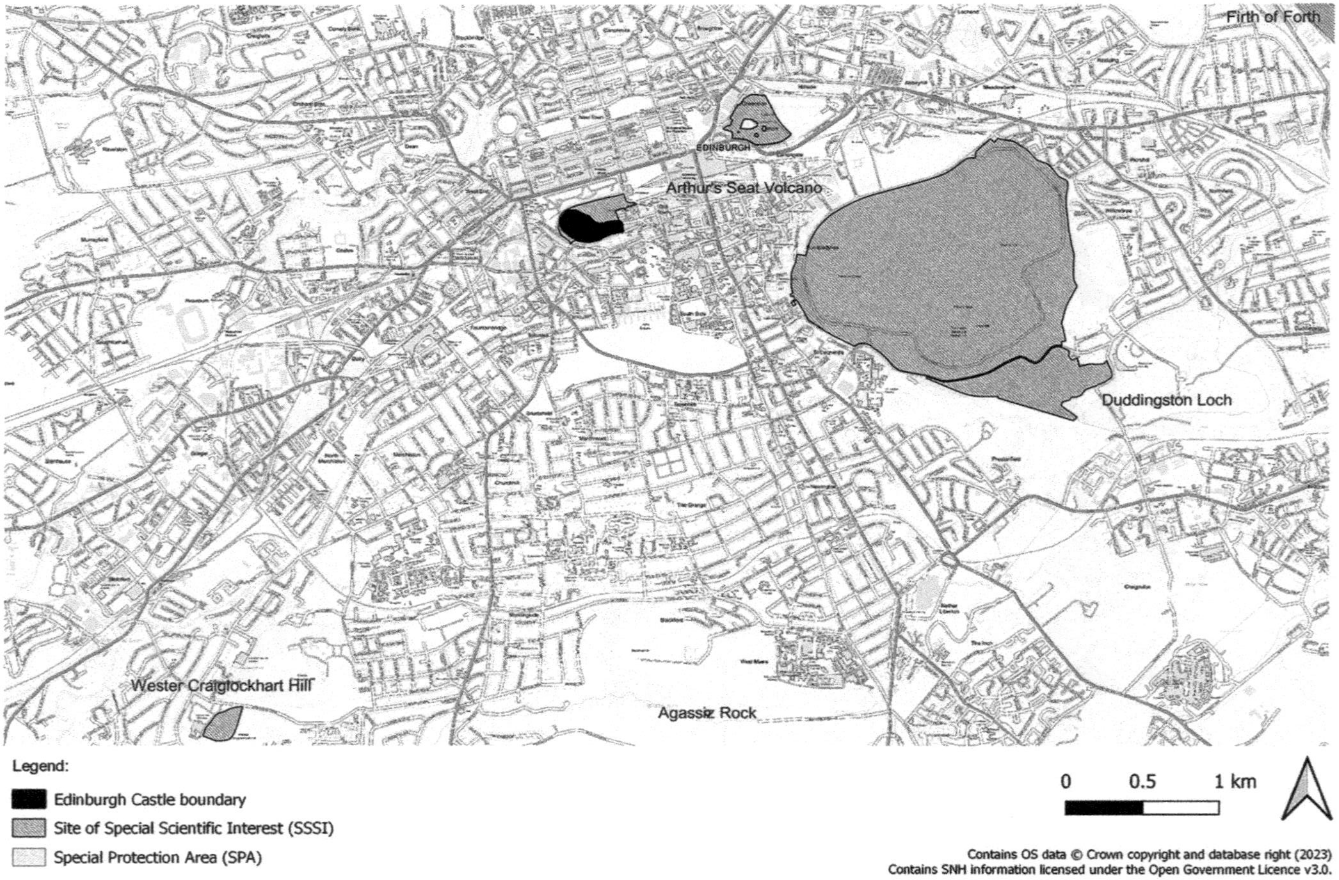

Figure 7.35 Designated sites map export

7.14   England

To download designated sites for England, visit the Natural England Open Data Geoportal web page and search for the dataset you want. At time of writing the url is: https://naturalengland-defra.opendata.arcgis.com/

Figure 7.36 Natural England Geoportal

Search for the dataset you want and click to open page.

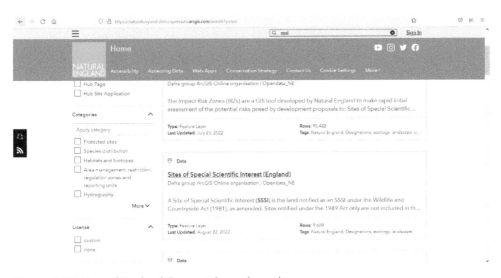

Figure 7.37 Natural England Geoportal search results

Click 'Download'.

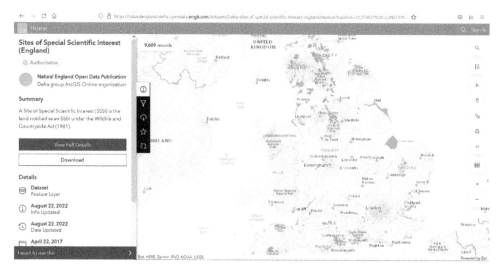

Figure 7.38 Download designated sites England

Wait for download and **unzip** and **save** in suitable location on **:C: drive**. **Load** into QGIS. **Zoom** to a desired scale.

OR

As we did in '6.4 Connecting to Aerial Imagery' instead of downloading the shapefiles we can connect to via a url.

These are hosted by ESRI again so we need to create a new ArcGIS REST Server connection.

At time of writing you find the url for setting up the connection by following the steps above for data download on the Natural England Geoportal, except this time clicking the '**View Full Details**' button instead of the 'Download' button on the data page (See Figure 4.28 Download designated sites England).

Scroll down and Open the 'View API Resources' submenu. Copy the url and paste into a document. At time of writing for SSSIs this is: https://services.arcgis.com/JJzESW51TqeY9uat/arcgis/rest/services/SSSI_England/FeatureServer/0

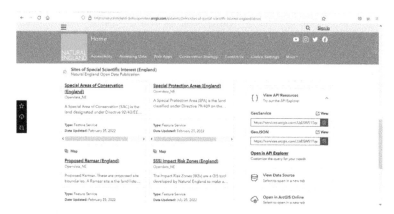

Figure 7.39 Dataset details and url

We want to set up a connection to all the data held on the services so we only need part of this url: https://services.arcgis.com/JJzESW51TqeY9uat/arcgis/rest/services

Back in QGIS, **Click** on '**Open Data Source Manager'** in the top toolbar.

Scroll down and Click the '**ArcGIS REST Server'** tab from the left side of the window.

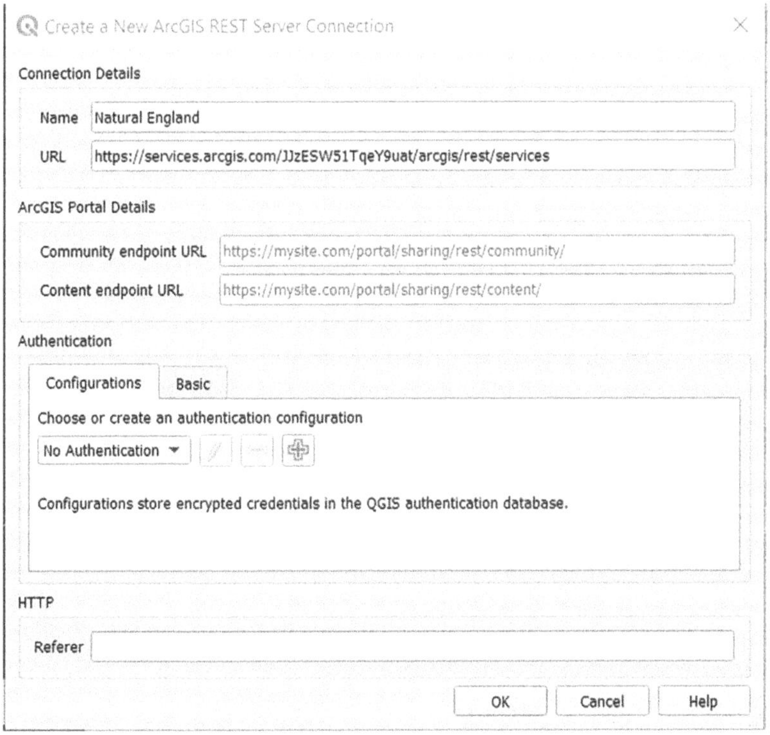

Figure 7.40 Create new connection for designated sites England

Type a name for the Connection, e.g. 'Natural England'.

Currently the url for this is:

https://services.arcgis.com/JJzESW51TqeY9uat/arcgis/rest/services/

Click 'OK'

Back in the **Data Source Manager** window **click** '**Connect'**

Select the layers then **click Add.**

In the Browser panel you should now see a huge list of layers when you open the Natural England dropdown under ArcGIS REST Servers. Scroll down in the layers panel to find and add in the layer you want. However, it is often quicker to load and easier to search within the Data Source Manager connection to find the layers you want to load in.

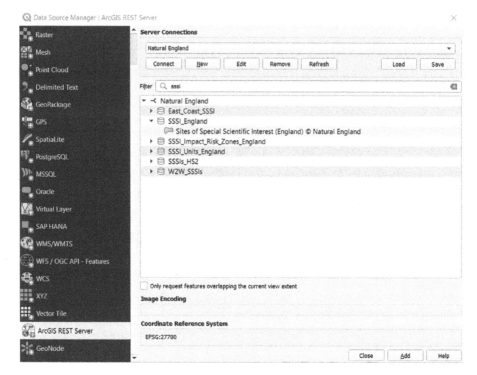

Figure 7.41 Add in dataset from Natural England connection

Do you notice a difference between the appearance of these layers and the layers we connected to for Protected Areas in Scotland?

....

Why do you think this is?

....

....

Follow your instructions in '7.7: My QGIS Instructions for... measuring and recording distances' to measure the approximate distance from your site to the nearest SSSI, SAC, SPA. If relevant to your site you might want to also add RAMSAR, National Nature Reserves and Local Nature Reserves to the map and measurements.

Follow the steps in '7.10 My QGIS Instructions for... creating a map using styles and legend'.

Look up the copyright statement on the website where you downloaded the data from and add it to the map.

7.15   Wales

To download designated sites for Wales, go to the Data Map Wales website and search for the dataset you want. At time of writing the url is: https://datamap.gov.wales/ Search in the search box for the dataset you require.

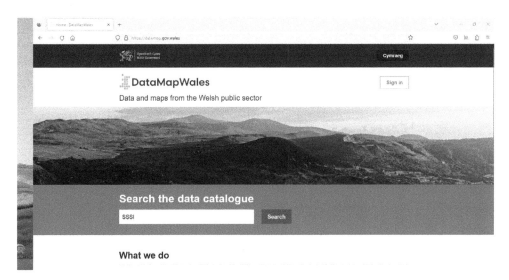

Figure 7.42 Data Map Wales

Click on the dataset you want to be taken to the dataset download page.

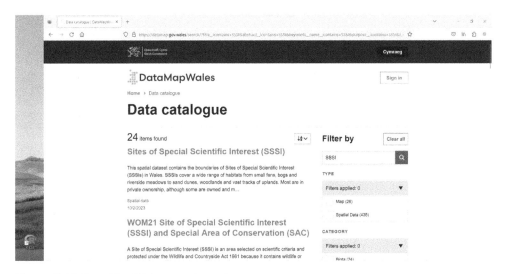

Figure 7.43 Data Map Wales search results

In the dataset page, **scroll down**, under '**Spatial data download**' **click** on the '**Select Spatial Download**' dropdown menu and **click** on '**Zipped Shapefile**'.

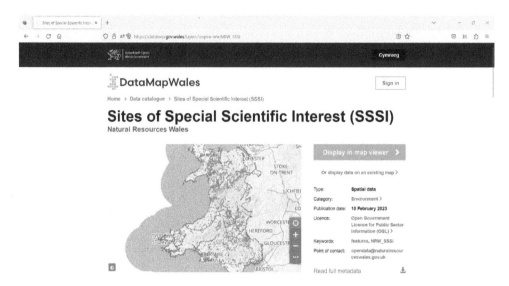

Figure 7.44 Data Map Wales download page up

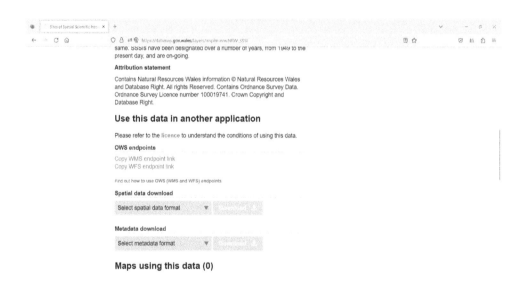

Figure 7.45 Data Map Wales download page down

Wait for download and **unzip** and **save** in suitable location on **C: drive**. **Load** into QGIS. **Zoom** to a desired scale.

OR

As we did in '6.4 Connecting to Aerial Imagery' instead of downloading the shapefiles we can connect to via a url.

At time of writing, you find the url for setting up the connection by following the above steps for downloading from the Data Map Wales webpage except this time **click** under 'OWS endpoints' you can choose to copy either the WMS or WFS endpoint or url.

The process for connecting to the WMS is the same as described above for the Nature Scot connection, except that you need to add in each designated site as a separate url or you can add in the WFS following the instructions below.

In the 'Data Source Manager' accessible from the top toolbar, we want to select '**WFS/OGC API Features**' tab.

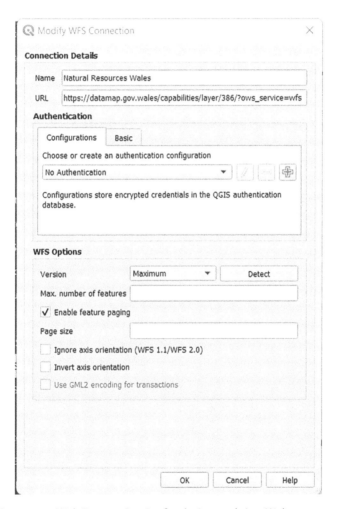

Figure 7.46 Connect to Web Feature Service for designated sites Wales

Type a name for the Connection e.g. 'Natural Resources Wales'
Currently the url:
http://lle.gov.wales/services/wfs/nrw
Click 'OK'
Back in the **Data Source Manager** window **click** 'Connect'.
Select the layer then **click Add.**

Do you notice a difference between the appearance of these layers and the layers we connected to for Protected Areas in Scotland?

....

Why do you think this is?

....

....

Follow your instructions in '7.7: My QGIS Instructions for... measuring and recording distances' to measure the approximate distance from your site to the nearest SSSI, SAC, SPA. If relevant to your site you might want to also add RAMSAR, National Nature Reserves and Local Nature Reserves to the map and measurements.

Follow the steps in '7.10 My QGIS Instructions for... creating a map using styles and legend'.

Look up the copyright statement on the website where you downloaded the data from and add it to the map.

7.16 Northern Ireland

To download designated sites for Northern Ireland, go to the Open Data NI webpage and search for the dataset you want.

At time of writing the website is:

https://www.opendatani.gov.uk/

Figure 7.47 Open Data NI

Scroll down and **Click** on the link to download the dataset you want.

Figure 7.48 Open Data NI search results

On the download page **click** on '**Download**' beside e.g. '**SHP Areas of Special Scientific Interest (ASSIs)**'

Figure 7.49 Open Data NI data download

Wait for download and **unzip** and **save** in suitable location on **C: drive**. **Load** into QGIS. **Zoom** to a desired scale.

OR

As we did in '6.4 Connecting to Aerial Imagery' instead of downloading the shapefiles we can connect to via a url.

At time of writing you find the url needed on to the Get Data page of the Environmental Protection Agency Geoportal website: https://gis.epa.ie/GetData/Connect

Scroll down and **Open** the 'OGC WMS service' submenu.
Copy the url.

In the 'Data Source Manager' accessible from the top toolbar, we want to select 'WMS/WMTS' tab.
Type a name for the Connection e.g. 'NI EPA'.

Paste the url into the url box, currently this is:

https://gis.epa.ie/geoserver/wms?service=WMS&request=getCapabilities&version=1.3.0

Figure 7.50 Create new connection for designated sites NI

Click 'OK'.

Back in the **Data Source Manager** window **click 'Connect'**.

Select the layers then **click Add.**

In the Browser panel you should now see a huge list of layers when you open the NI EPA dropdown under WMS/WMTS Features. Scroll down in the layers panel to find and add in the layer you want. However, it is often quicker to load and easier to search within the Data Source Manager connection to find the layers you want to load in.

There does not currently appear to be a WFS available for Northern Ireland.

Follow your instructions in '7.7: My QGIS Instructions for... measuring and recording distances' to measure the approximate distance from your site to the nearest SSSI, SAC, SPA. If relevant to your site you might want to also add RAMSAR, National Nature Reserves and Local Nature Reserves to the map and measurements.

Follow the steps in '7.10 My QGIS Instructions for... creating a map using styles and legend'.

Look up the copyright statement on the website where you downloaded the data from and add it to the map.

Congratulations! You have now made a designated sites map for your site!

7.17 My Instructions for... downloading vector data from government sources and government data copyright

What are the four steps for downloading vector data from a government website?

1. ....

2. ....

3. ....

4. ....

What are the six steps for setting up a new connection to a web map service from a government website?

1. ....

2. ....

3. ....

4. ....

5. ....

6. ....

What is the copyright statement for NatureScot data?

....

What is the copyright statement for Natural England data?

....

....

What is the copyright statement for Welsh Government/Natural Resources Wales data?

....

....

What is the copyright statement for Environmental Protection Agency Northern Ireland data?

....

....

Connecting to Ordinance Survey data

Unlike our survey maps for taking into the field as printed maps or for a quick check on where designated sites are in relation to our site, in order to produce maps for our reports we may wish to access higher resolution raster basemaps or vector data.

In addition to the OS Open Data downloads (see 5.8 'Downloading basemaps from OS Open Data') you can also connect to this data as a web map/feature service.

To do so visit the **OS Data Hub website** and **sign up** for an account. At time of writing this is: https://osdatahub.os.uk/plans

You can access all the open data available for download and pay to access other services such as OS Mastermap. OS Mastermap provides high-resolution vector data of the UK down to field boundary level. If you work in the public sector your organization may already have access to this resource.

Once your account is set up you can **generate APIs** which are the urls needed to connect to the web map/feature service. Instructions and help are provided by Ordinance Survey to help you with this process, so are not covered here.

However, once in QGIS the methodology for setting up the connection follows that detailed in 'Chapter 7. Designated Sites map – Accessing Designated Sites data'.

Add in connections in **Data Source Manager** based on the type of data, for instance:

For **raster** tiles use: **WMS/WMTS**

For **vector** tiles use: **Vector Tile**

For **features** use: **WFS/OGC API – Features**

8. Desk-study maps

Before going into the field, you need to know what has been recorded on the site previously; the best way to see this is to map existing records. For this exercise we will be looking at records of beaver along the Tay river complex in south-east Scotland. We have a spreadsheet of biological records from Nature Scot that we want to import into QGIS and create a map for our report from.

You will learn:

- How to create and import point vector data from a spreadsheet.
- What is a plugin?
- How to import data from NBN.
- How to troubleshoot in data creation.
- How to produce desk species maps from created data.

Creating point data: importing from a spreadsheet

Set up a new project called 'Tay Beavers' following '7.3 My QGIS Instructions for... setting up a Project'.

Add the **OpenStreetMap** basemap layer.

Change CRS to 'EPSG:27700 – OSGB36 / British National Grid'.

The beaver records we want to import are stored in: "Beaver_Tay_SNH.csv"

What is a CSV file?

A CSV file is a simple table file that can be opened in Excel. CSV stands for 'comma separated values' and refers to how the data is stored as columns and rows separated by commas. To import spreadsheets into QGIS they need to be saved as .csv files not .xlsx files as is the default when saving in Excel.

Open "Beaver_Tay_SNH_NBN_OGL.csv" **using MS Excel**

What information is contained in the first row?

....

What data is contained in the OSGR column?

....

What is the difference between the data in the OSGR column and the two proceeding columns?

....

....

....

Ordnance Survey National Grid (OSGR), Eastings and Northings

Ordnance Survey National Grid splits the UK into 55 squares of 100 × 100km described by two letters. Coordinates in British National Grid can be written in Ordnance Survey National Grid which have two letters followed by numbers. British National Grid coordinates can also be written numerically as Eastings and Northings.

8.1   Importing Ordnance Survey National Grid Coordinates

In order to import the spreadsheet using the OSGR coordinates we need to use the TomBio Plugin.

What is a plugin?

A plugin is an additional tool for QGIS that allows us to do complicated mapping tasks more quickly, like a shortcut. There are plugins available for a number of biological mapping tasks and we will be having a look at some of these in this workbook.

To do so we first need to install the plugin.

In the top menu **Click** on '**Plugins**' then '**Manage and Install Plugins**'.

Figure 8.1 Manage and install plugins

When the window opens, **select** the 'All' tab from the left side and **type** 'TomBio' in the search bar.

Click 'Install plugin' and wait for it to install.

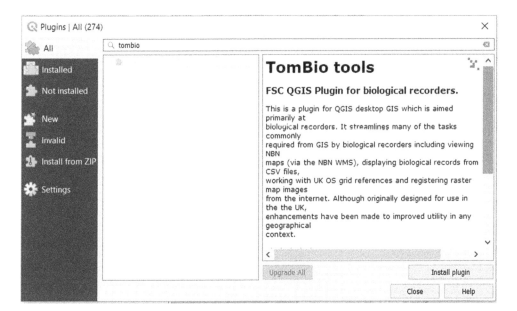

Figure 8.2 TomBio tools plugin

8.2   My QGIS Instructions for... installing plugins

What are the four steps for installing a plugin?

1.   ....

2.   ....

3.   ....

4.   ....

A new toolbar should appear in the top toolbar, if not restart QGIS.

Figure 8.3 TomBio toolbar

If it does not, right-click on the top toolbar and tick the box beside 'TomBio Toolbar' to turn this on.

On the TomBio Toolbar **click** the 'Biological records tool' button.

Figure 8.4 Biological Records Tool

Under the 'Data specification' tab, Under 'Source definition' select 'Create source form CSV file' from the dropdown menu.

Click on the three dots to the right of the dropdown menu and browse to the "Beaver_Tay_SNH_NBN_OGL.csv" file.

To import the coordinates in British National Grid format, select the column containing the grid reference 'OSGR' from the OS Grid Ref Column pull down menu. Alternatively, you can import the coordinates as 'Eastings' and 'Northings' using the 'X' and 'Y' dropdown menus respectively.

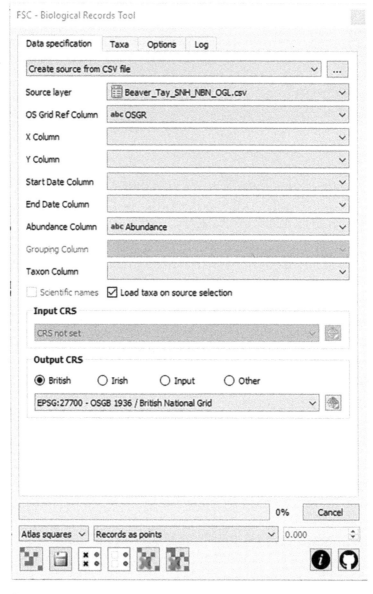

Figure 8.5 Beavers csv TomBio import

Then click **create map layer**

Figure 8.6 Map records

Once the new layer has been added to the map legend, convert it to a shapefile. **Right-click** on the layer and select **Export** then **Save features as...**

Figure 8.7 Export

Select **ESRI Shapefile** as the format and ensure that the **CRS** is **British National grid**. Select **Add saved file to map**. Browse to a suitable folder and save the new file. Remove the original file from the map legend; right-click on the layer and select **remove**.

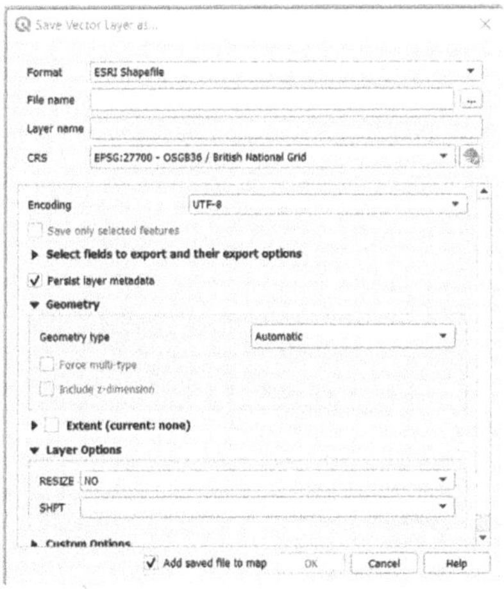

Figure 8.8 Save as shapefile

Why do we need to convert to a shapefile?

....

....

....

Create a spatial index for the file:

In the top menu click 'Vector', then 'Data Management Tools' then 'Create Spatial index'.

Select the 'Beaver_Tay_SNH_NBN_OGL.csv' layer from the dropdown and click '**Run**'

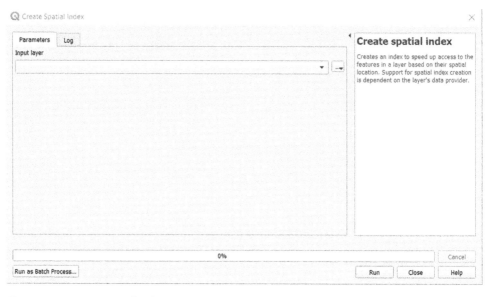

Figure 8.9 Create spatial index

Why create a spatial index file?

Creating a spatial index file ensures help large shapefiles work quickly and insures they plot in the right place on other computers and in other software. It is good practice to create a spatial index every time you create a shapefile, as it saves trouble later!

8.3 My QGIS Instructions for... importing Ordnance Survey National Grid Coordinates

What are the six steps for importing a spreadsheet using the TomBio Toolbar Biological records tool?

1. ....

2. ....

3. ....

4. ....

5. ....

6. ....

8.4 Styling point data

We are now going to style our data.

To change the symbology of a layer, **double-click** a layer in the layers panel (or **right-click** and select '**Properties**'). This opens the layer properties.

Choose the symbology tab from the left side of the pop-up box.

Figure 8.10 Symbology with point data

Click on 'Simple marker' in the top box.

Choose 'Simple marker'.

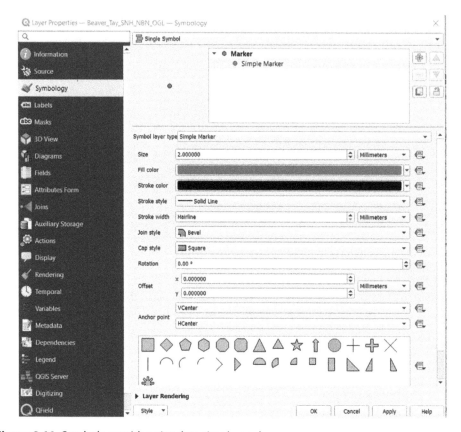

Figure 8.11 Symbology with point data simple marker

Change the colour by choosing from the palette of standard colours.

Change 'fill style, stroke color' from the drop-down menus as for the polygons layers. **Click 'OK'** to close the window.

Save project – This will save all changes.

8.5   Desk-study map from a spreadsheet

Follow '7.10 My QGIS Instructions for... creating a map with styles and legend'

Give your Layout the title of **"Tay Beavers Desk-study map".**

What is the copyright statement for the basemap?

....

Add copyright statement for the beaver records: "NatureScot (2019). Survey of distribution of beavers in Tayside in 2012. Occurrence dataset on the NBN Atlas."

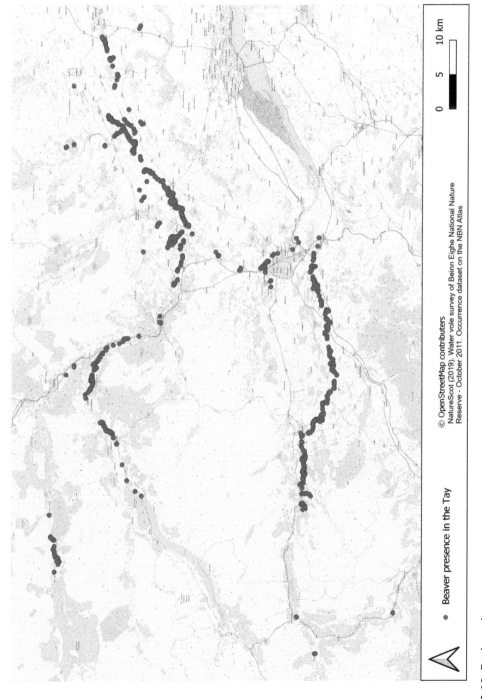

• Beaver presence in the Tay

© OpenStreetMap contributers
NatureScot (2019), Water vole survey of Beinn Eighe National Nature
Reserve - October 2011. Occurrence dataset on the NBN Atlas

0   5   10 km

Figure 8.12 Desk-study map export

Creating point data: importing data from NBN

8.6 Downloading data from NBN

The National Biodiversity Network (NBN) is a great resource for species information. However, most of it cannot be used commercially.

NBN data copyright

For working with NBN data we need to first check the licence conditions on the NBN website and request access where required. Data on the NBN under an Open Government Licence (OGL) can be used commercially. Most data on the NBN is under a Creative Commons Licence (CC) and cannot be used commercially without permission or payment.

Go to the **NBN Atlas website** and **log in** or **create** a log in. At time of writing this is: https://nbnatlas.org/

Search for the "Squirrel Collation Records" dataset from Natural Resources Wales and look up the licence. At time of writing the webpage for this is: https://registry. nbnatlas.org/public/show/dr899

Download the data by first clicking on 'View Records' then on the 'Overview and Downloads' tab then 'Download'.

In Step 1 **select** the download of the Occurrence Records.

In Step 2 **select** the appropriate reason for records download, this will usually be 'commercial' in your work but in this case is 'education'. Tick to accept the licence conditions.

Download the data from the link sent to your email.

Unzip the folder, **save** on the C: drive and **Open** the 'records-2023-07-17.csv' in Excel. At the right end of the spreadsheet there are 4 OSGR columns.

What do the data show?

....

Coordinate accuracy

On the British National Grid, a grid reference can be written in Ordnance Survey National Grid format which has two letters followed by varying amounts of numbers and/or letters dependent on the accuracy. British National Grid coordinates can also be written numerically as Eastings and Northings always with six digits for Eastings and six/seven digits for Northings. For example, Edinburgh castle locations could be given to 10km accuracy/hectad NT 27 or 320000, 670000; 1km accuracy/tetrad NT 25 73 or 325000, 673000; 100m accuracy NT 25139 73486 or 325139, 673486.

Follow your '8.2 My QGIS Instructions for... importing Ordnance Survey National Grid Coordinates'. **Choose** 'Coordinates'. **Select** 'Scientific name' from the Taxon column dropdown menu. Then 'Order' from the 'Grouping' column and **tick** the squirrel species from the 'Taxa' tab.

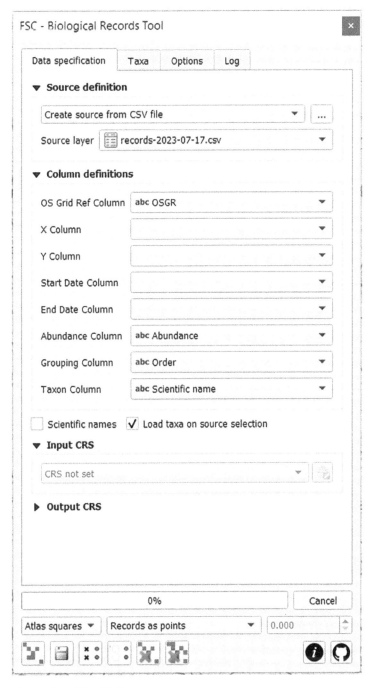

Figure 8.13 Squirrel csv TomBio import

Figure 8.14 Squirrel csv TomBio import taxa tab

Save the features as **shapefile**. Create a spatial reference.

Repeat the process again, but this time **select '10km atlas (hectads)'** from the 'Records as points' dropdown menu.

Describe the output on the map:

....

Save the features as **shapefile**. Create a spatial reference.

Compare the Attribute table for the shapefile imported by mapping 'Records as points' with Attribute table of the shapefile imported by mapping 'Atlas squares' and the original CSV file, what do you notice? Why do you think this is?

....

....

Which method would be useful for which circumstances?

Records as points import ....

....

Atlas squares import ....

....

Which do you think would be most applicable to your work and why?

....

....

Of course, you can import using both methods and save the respective shapefiles as 'Squirrel records points' and 'Atlas squirrel records', for instance.

8.7   My QGIS Instructions for... downloading data from the NBN

What are the six steps for downloading data from the NBN?

1. ....

2. ....

3. ....

4. ....

5. ....

6. ....

Connecting to NBN data

Once we have found OGL data or acquired access to the NBN dataset we want, we can then connect to it directly via the TomBio Toolbar.

Go to **TomBio Toolbar** and find the **NBN Atlas Tool**

Figure 8.15 NBN Atlas Tool

Select the 'Filter' tab then the side '**Species**' tab and search for '**Squirrel**' and select '**Sciurus**' with a **double left-click** from the bottom of the **Taxon** tree.

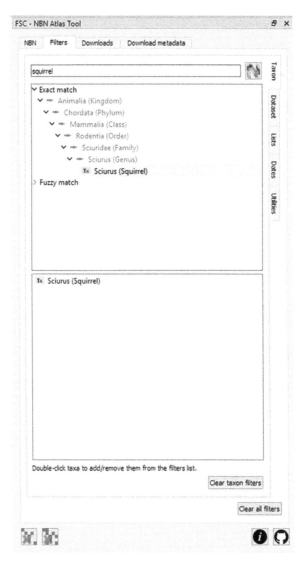

Figure 8.16 NBN Atlas Tool Filter Taxon

Select the 'Dataset' tab and select 'Natural Resources Wales' then 'Squirrel Collation for Wales'.

Figure 8.17 NBN Atlas Tool Filter Dataset

Select the 'NBN' tab and select 'Atlas squares' from the top dropdown menu. Then click 'Map'.

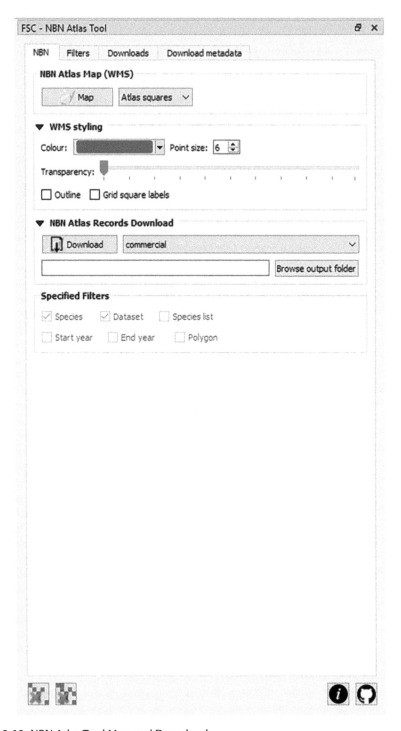

Figure 8.18 NBN Atlas Tool Map and Download

What scale are the Atlas squares and why?

....

What do you notice about the layer when it appears in the Layers panel?

....

Why do you think this is?

....

What would happen if we saved the features in this layer?

....

In order to categorize the layer, we need it to be a vector dataset.

To do so we can download the data as we did before or by using the download function within the NBN Atlas Tool.

We have the option to either download 'Record points' or 'Atlas squares', which are 10km atlas (hectads). Choose 'Atlas squares'. **Select** your reason for download – this will usually be 'commercial' in your work but in this case is 'education'. **Use** 'Browse output folder' to select where to download file to. The **'Downloads'** tab of the NBN will automatically be shown if download is successful.

Figure 8.19 Squirrel csv download via NBN Atlas Tool

Double left-click on the csv file shown in the 'Downloads' tab and the Biological records tool will automatically pop-up. **Follow** the steps in '8.2 My QGIS Instructions for... importing Ordnance Survey National Grid Coordinates'.

8.8   My QGIS Instructions for... connecting to data from NBN

What are the seven steps for connecting to data from NBN?

1.   ....

2.   ....

3.   ....

4.   ....

5.   ....

6.   ....

7.   ....

Which of the methods for downloading and importing data from the NBN do you prefer and why?

....

Direct data connection to the NBN

Connecting directly to the NBN via the NBN Atlas Tool and importing via the Biological records tool is perhaps more straight forward but without checking the NBN website first you may not get full access to the data where you may have done had you asked permission or paid for the dataset. Note: Sometimes this direct access to the NBN may not work, for instance because the site is under maintenance.

8.9   Categorized polygon symbology

Double-click on the Squirrel records polygon layer in the left layers panel or **right-click** and select 'Properties'.

Choose the '**Symbology**' tab.

At the top where 'Single symbol' is automatically shown, click on the dropdown menu arrow to the right and select '**Categorized**'.

Under 'Column' choose the '**Taxa**' to categorize by this field from the shapefile.

Click 'Classify'. Arbitrary colours will be applied to each species grouping. **Double-click** on each symbol and choose the style you would like for each. **Choose** a different colour for each species and add some transparency.

8.10   Desk-study map from NBN data

Export a desk-study map of the squirrel records as 10km grid squares, showing the overlap of species.

Legend:

Squirrel records of Wales 10km Atlas

Both species present

Red squirrel (Sciurus vulgaris)

Grey squirrel (Sciurus carolinensis)

0   25   50 km

Contains OS data © Crown copyright and database right (2023)

Records provided by Natural Resources Wales, accessed through NBN Atlas website
Natural Resources Wales. (2023) Squirrel Collation for Wales. Occurrence dataset accessed through the NBNAtlas

Figure 8.20 NBN Desk-study export

Follow '8.2 My QGIS Instructions for... importing Ordnance Survey National Grid Coordinates' or '8.9 My QGIS Instructions for... importing data from NBN' and '8.2 My QGIS Instructions for... importing Ordnance Survey National Grid Coordinates'.

Try to remember the steps in '7.10 My QGIS Instructions for... creating a map with styles and legend' without looking at them to produce a desk-study map of your records.

What is the copyright statement and where can it be found?

....

Create your own desk-study map by downloading data from the NBN or importing data from the NBN via TomBio tools.

Follow '8.2 My QGIS Instructions for... importing Ordnance Survey National Grid Coordinates' or '8.9 My QGIS Instructions for... importing data from NBN' and '8.2 My QGIS Instructions for... importing Ordnance Survey National Grid Coordinates'.

Try to remember the steps in '7.10 My QGIS Instructions for... creating a map with styles and legend' without looking at them to produce a desk-study map of your records.

Fantastic! You have now made a desk-study map for your site using NBN data!

9. Protected species survey map

You have returned from the field with protected species target notes and a GPS containing location data. Your protected species target notes have been typed up into a spreadsheet and you have uploaded the GPS data to the computer. You want to get this data onto QGIS and produce a protected species map for your report.

You will learn:

- More functions of QGIS Desktop interface
- How to import point vector data from GPS field survey data
- How to troubleshoot in data creation
- More functions of QGIS Print layout
- How to produce a protected species map from created data

Creating point data: importing from a GPS

In order to import data from a survey using coordinates recorded using a GPS into QGIS, the file needs to be saved as a GPX. Software and online tools are available for this for instance GPSBabel. Here we will use an example GPX file to import into QGIS.

Why import GPS data?

We could choose to type up our coordinate data and target note information from our GPS and/or notebook and into a spreadsheet and then import this into QGIS using '8.2 My QGIS Instructions for... importing Ordnance Survey National Grid Coordinates'. The problem with this is the increased number of steps and need for transcription increases not only the amount of effort but the chance for errors to creep in.

Create a QGIS project called: "**Beinn Eighe Water Voles.qgz**"

Instead of CRS EPSG:27700 – OSGB36 / British National Grid set the CRS to ESPG:4326 – WG84.

Add in basemap of your choice.

We are going to import 'Water Voles Beinn Eighe.gpx' to QGIS.

9.1   Importing GPX files

Open the **Data Source Manager** and click the **'GPS'** tab.

Use the '...' button to select the GPX file.

Note: You can import multiple GPX files using the Batch GPS Importer Plugin.

Leave **Waypoints** ticked but **untick Tracks and Routes** as there are none in the file.

Figure 9.1 Data Source Manager GPS

Save as a shapefile with a sensible name and **Create** a spatial index for the file. **Change the CRS to 'EPSG:27700 – OSGB36 / British National Grid'.**

Why did we set CRS ESPG:4326 – WG84 instead of EPSG:27700 – OSGB36/ British National Grid for importing the GPS points?

....

....

9.2 My QGIS Instructions for... importing from a GPS

What are the six steps for importing from a GPS?

1. ....

2. ....

3. ....

4. ....

5. ....

6. ....

9.3 Troubleshooting with GPS

Why might points taken from a GPS not plot on the map exactly where you expect?

....

....

....

....

Describe the ways you might change where a point plots on the map.

....

....

....

....

....

....

....

....

9.4 Categorized symbology

We are going to make a Protected Species survey map using the 'Beinn Eighe Water Voles' project you have created.

Open the Attribute table of the Water Vole shapefile, **right-click** then 'Open Attribute table'.

What information is stored in the 'Description' column?

....

How can we simplify this information in order to map it?

....

....

Create a new field in the Attribute table called '**Category**'. (See Chapter 7.5: Adding an Attribute field).

This time we want to add a field of 'Text' type (string) of the length (in characters) that we wish to insert; we can add in up to the maximum, which is '250'. Allowing 250 characters will enable you to enter as long an amount of text as possible. If we leave it as the default 10, then we end up with not enough characters to describe the categories.

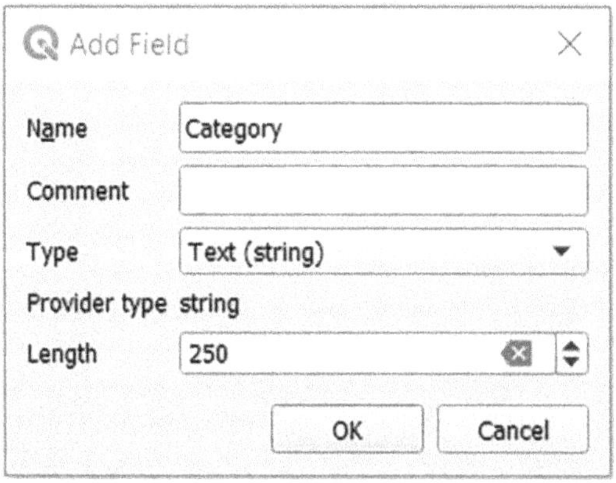

Figure 9.2 Add text field

Decide on 4 or 5 named categories that simplify the information in the '**Description**' column for each record. **Type** these categories in the '**Category**' column.

In order to get this information onto the map, we can categorize the points by changing their colour.

Use the map navigation tools (Pan, zoom, zoom to layer etc.) so you can see all the points layers in your map view.

We are going to categorize the Water vole survey points.

Double-click the layer in the left layers panel or **right-click** and select 'Properties'.

Choose the '**Symbology**' tab.

At the top where 'Single symbol' is automatically shown, click on the dropdown menu arrow to the right and select '**Categorized**'.

In '**Value**' dropdown choose the '**Category**' to categorize by this field from the shapefile.

Click '**Classify**'. Arbitrary styles will be applied to each symbol. **Double-click** on each symbol and choose the style you would like for each. **Choose** a different colour for each description.

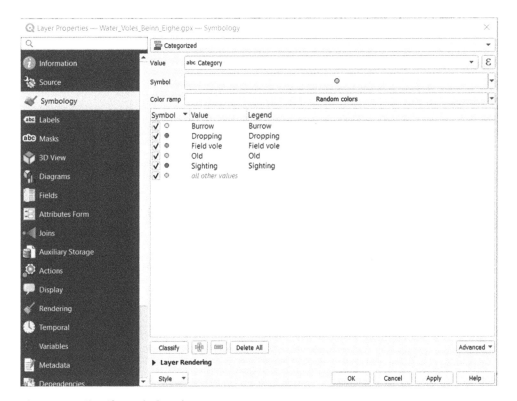

Figure 9.3 Classify symbology by category

Still in Layer Properties, **Click** on the '**Labels**' tab.

Select '**Single labels**' and select '**Name**'.

Why is it useful to categorize by colour and to label them on the map?

....

....

9.5   Protected species survey map

To give more context to the map, you may like to find and download the OS Open data contours dataset OS Terrain® 50. Or you can use the contour line shapefiles provided: NH05, NH06, NH95, NH96.

Follow the steps in '7.10 My QGIS Instructions for... creating a map using styles and legend'.

Give your Layout the title of "Beinn Eighe Protected Species map".

Use the copyright statement: "NatureScot (2019). Water vole survey of Beinn Eighe National Nature Reserve – October 2011. Occurrence dataset on the NBN Atlas."

Alter the way the subgroup font and item fonts appear in the legend using the "Fonts" dropdown in 'Item properties' for the legend. For instance, you may want the font size of the subgroup heading 'Category' to be larger than the items or underlined.

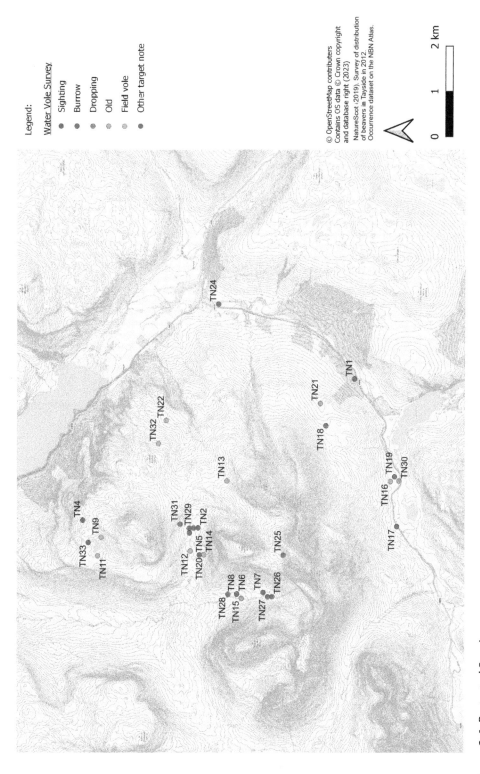

Figure 9.4 Protected Species map export

Fabulous! You have now made a Protected species survey map using GPS data!

9.6 My QGIS Instructions for... creating a map using categorized styles and legend

Describe in your own words how to use categorized symbology to style a vector file.

....

....

....

List the eight steps you used to produce your Protected Species map:

1. ....

2. ....

3. ....

4. ....

5. ....

6. ....

7. ....

Describe in your own words how to add and alter a legend in the Layout with categorized vectors.

....

....

....

Create your own protected species survey map from your own GPX file from a field survey. **Follow** '7.3 My QGIS Instructions for... setting up a Project' and '9.2 My QGIS Instructions for... importing from a GPS'. **Follow** the steps in '9.6 My QGIS Instructions for... creating a map using categorized styles and legend' to produce a protected species survey map using your records.

Good work! You have now made a Protected species survey map for your site using your GPS data!

10. Georeferencing maps

You have returned from the field with maps with field notes on or you have been given a paper map by a landowner. You have scanned in this map image and want to use it as a base for digitizing a habitat map.

You will learn:

- How to prepare field maps for use in QGIS
- How to georeference a field map with grid-lines
- How to georeference a field map without grid-lines

10.1   Georeferencing using OS grid-lines

Create a new project called "**Georeferencing**"

Add in Raster layer '**NS59NE.tif**'

We are going to use the **Georeferencer** tool to import a scanned in map and align it with the basemap.

Preparing images for Georeferencing

Images to be georeferenced need to be saved as jpeg or jpg format. Try to flatten your field map as best as possible and scan it in as straight as possible within the scanner. If your scanner does not allow you to save directly to jpeg or jpg, use an image manipulation program or take a screenshot to save a copy of the image as *.jpeg or *.jpg once it is scanned in.

Photographs of images taken on a mobile device can be used for georeferencing; however, these are often warped as it is very difficult to take a photograph straight into an image. If you only have a photograph of a map available, you will need to adjust the Transformation Settings to see what gives the best fit.

QGIS Versions 3.26 and above: the **Georeferencer** tool can be found in the **Layers** menu.

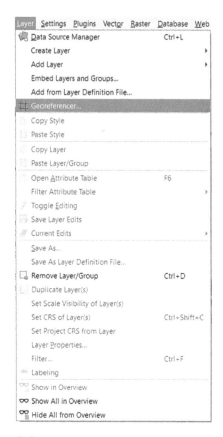

Figure 10.1 Georeferencer in Layer menu

QGIS Versions 3.24 and below: the **Georeferencer** tool can be found in the **Raster** menu.

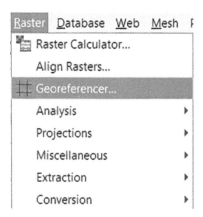

Figure 10.2 Georeferencer in Raster menu

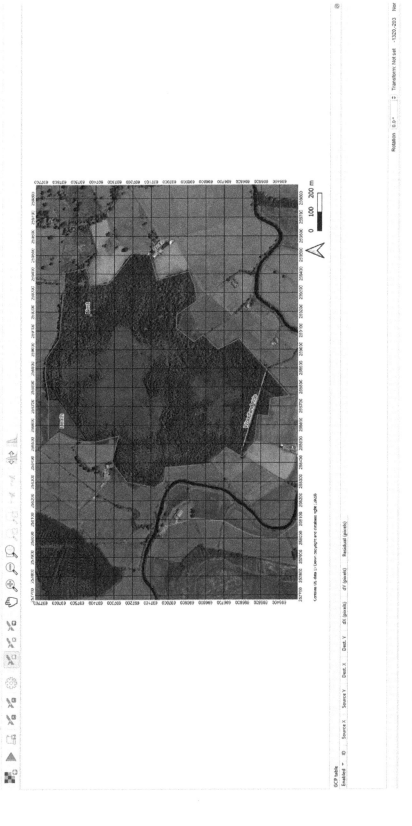

Figure 10.3 Load in grid map

In the **Georeferencer** window, **Click** on the **Open raster** button and **Open** 'Collymoon Sketch grid.jpg' in the Georeference plugin window.

We are going to add control points to the image to tell QGIS the coordinates from our basemap.

Control points

Control points are how we tell the image to 'attach' to the basemap. Aim to use points at all four corners of the map. For maps with grid-lines we can type in the coordinates written down the side of our grid. Look for intersections of roads, corners of buildings that you can find on the image you are georeferencing and in the basemap to which you are georeferencing. This takes some practice.

Use the mouse scroll and 'Pan' or the 'Zoom In' button to **zoom in** on the top left corner of the image in the Georeferencer window.

Click on the 'Add point' button.

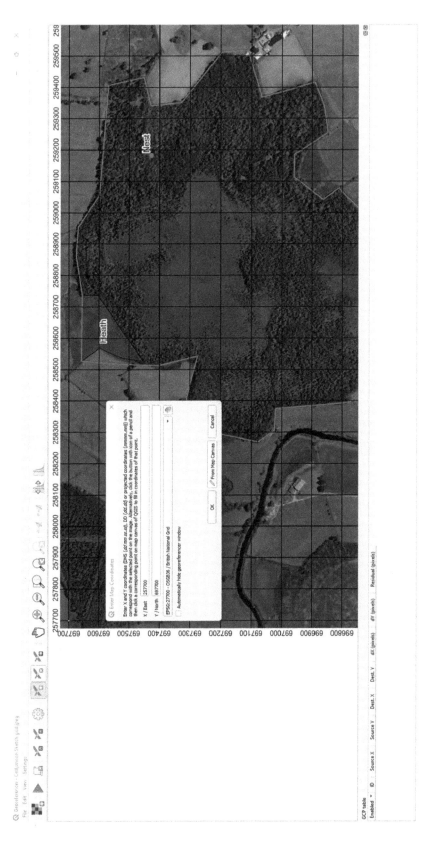

Figure 10.4 Add point and enter map coordinates

Line the cursor cross up with the very top left intersection of the grid-lines and **click** to create a control point.

You will see a small point has appeared where you clicked.

Type in the 'X/East' coordinate by reading off the coordinate from the top left of the image.

In this case '257700'.

Type in the 'Y/North' coordinate by reading off the coordinate from the left top of the image.

In this case '697700'.

Select 'OSGB36 / British National Grid EPSG 27700' for the Coordinate Reference System (CRS).

Untick 'Automatically hide georeferencer window'.

As you add each point what do you notice?

....

You can click on the points in the table below to select or delete them.

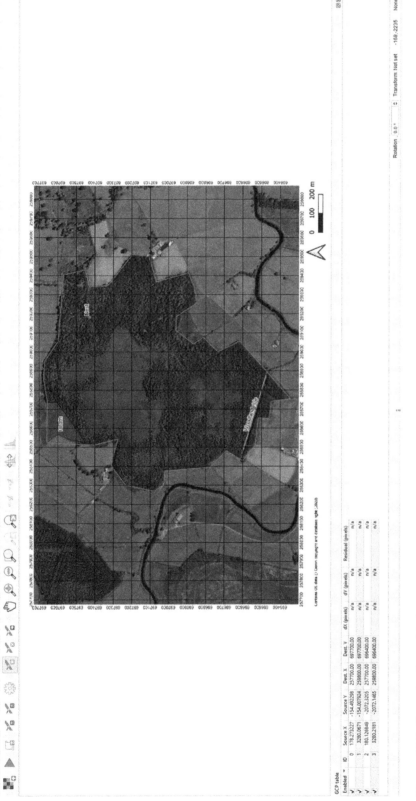

Figure 10.5 Control points added

Repeat this at the right top, left bottom and right bottom corners of the map, adding control points at X 259800, Y 697700; X 257700, Y 696400; and X 259800, Y 696400 respectively.

Click on the 'Transformation Settings…' button.

Change the Transformation type to 'Polynomial 1'.

Specify the file name and location of the output raster. Name the file with a different name to the input e.g. "Collymoon Sketch grid georeferenced.tif".

Set the target CRS to the same projection as the reference layer – British National Grid.

Leave re-sampling on the default 'Nearest Neighbour'.

Tick the 'Save GCP points' and the 'Load in project' when done.

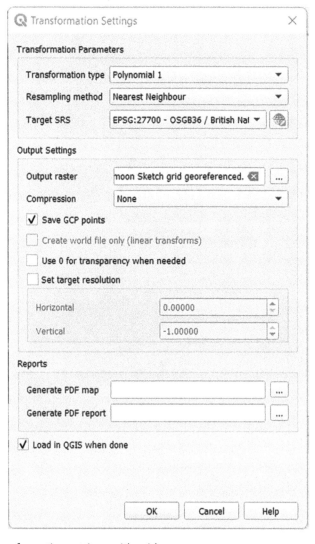

Figure 10.6 Transformation settings with grid

In the **Georeferencer** window, **Click 'Start Georeferencing'** button.

This complete the process and adds the image onto the map. **Close** the Georeferencer window using the 'X' in the top right of the window.

To check the accuracy of the georeferencing **right-click** on the newly added raster layer in the layers panel and open the **'Properties'** window.

Select the **'Transparency'** tab.

Use the slider to reduce the **'Global Opacity'** to 50%.

Figure 10.7 Making the Georeferenced image transparent

You now see how well the georeferenced image matches the basemap.

If necessary repeat the above in order to get a better fit.

The example result in the figure below shows the OS map image with grid-lines georeferenced to the OSM online basemap.

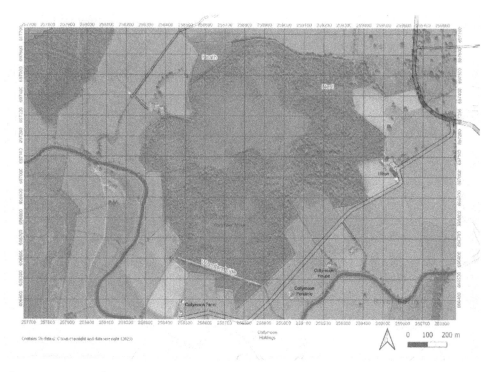

Figure 10.8 Georeferenced grid result with OSM basemap

Why do the images not exactly match up?

....

....

....

Basemaps for georeferencing

Ideally, we would georeference to the same basemap as the map used. However, this is not always possible. It is not too much of a problem as we would not display the georeferenced image on a printed map. Georeferenced images are best used to locate where to digitize features from survey maps. As accuracy of manual georeferencing in this way is limited, care is needed in digitizing from georeferenced images. Digitizing is covered in Chapter 11: Habitat Survey Maps.

Open the folder where you saved the georeferenced image.

What do you notice?

....

10.2   Georeferencing without OS grid-lines

In your **Georeferencing** project, make sure 'NS59NE.tif' is loaded and ticked in the layers panel.

Open 'Collymoon Sketch no grid.jpg' in the **Georeferencer** window.

We are going to add control points to the image to tell QGIS the coordinates from our basemap.

This time, we do not have grid-lines to help so we need to pick intersections of lines from the map image and basemap to georeference too. This can be quite tricky to master!

As before **Click** on 'Add Point' in the **Georeferencer** toolbar.

Select a point on the image then **Click** 'from map canvas' which will take you through to the basemap.

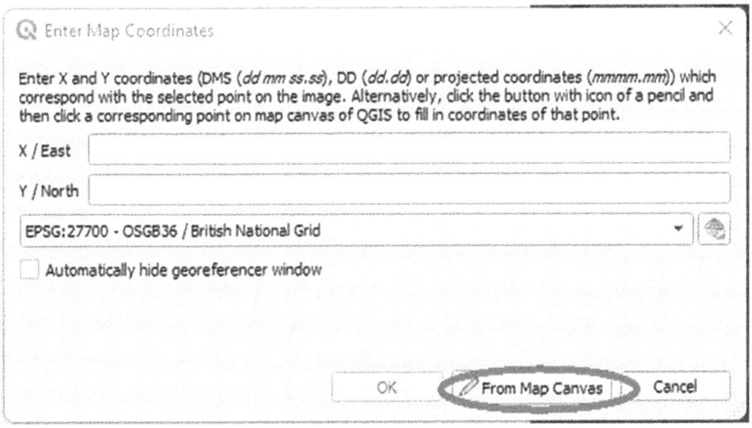

Figure 10.9 From map canvas

Select the same point on the main map canvas.

Add coordinates at the same positions as in the figure below.

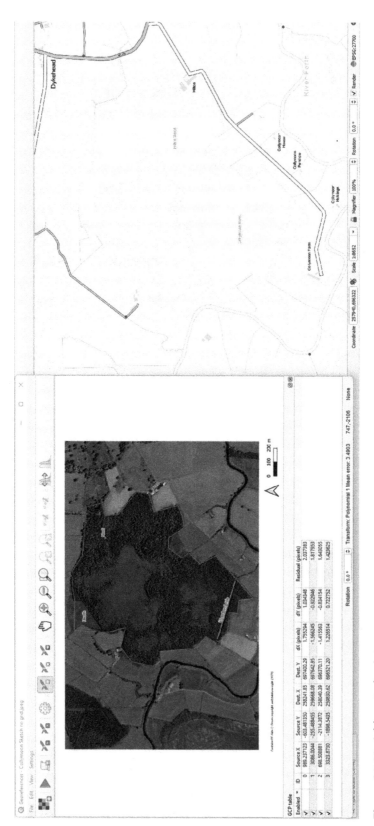

Figure 10.10 Add control points

Click on the points in the table below to select and move or delete.

Once you have enough points, go to '**Transformation settings**'.

Choose '**Thin Plate Spline**' for the transformation type.

Set the target CRS to the same projection as the reference layer – British National Grid. Leave re-sampling on the default 'Nearest neighbour'.

Name the file with a different name.

Tick the '**Load in QGIS**' box when done.

In the **Georeferencer** window, **Click** the green arrow to start georeferencing.

Transformation settings

These allow you to change the way the image is handled by the Georeferencer. Polynomial 1 is useful for flat images with grid-lines. Thin Plate Spline is useful for images without grid-lines. Images can be warped if field maps get wet or are folded, or if the image is from a photograph rather than a scanned source. It is preferable to avoid georeferencing a warped image by flatting and scanning in maps. If you need to georeference a warped map, changing the Transformation settings can improve the georeferencing accuracy.

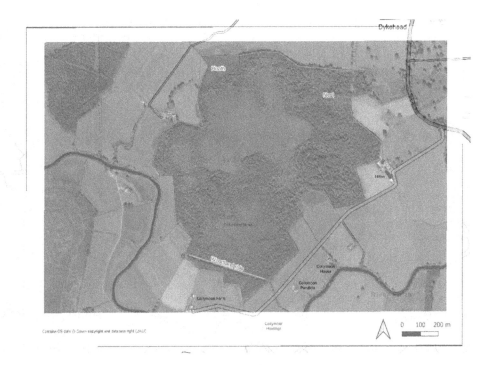

Figure 10.11 Georeferenced no grid result with OS basemap

How do we check if our georeferencing is accurate?

....

If necessary repeat the above in order to get a better fit.

10.3   My QGIS Instructions for... georeferencing an image with grid-lines

What are the six steps for georeferencing an image with grid-lines?

1. ....
2. ....
3. ....
4. ....
5. ....
6. ....

Create your own georeferenced map with grid-lines if you have one of a site you are working on by scanning it in and saving it as a .jpg then following the rest of your steps in *'10.3 My QGIS Instructions for... georeferencing a map image with grid-lines'*.

Great! You have now georeferenced your map image with grid-lines!

10.4   My QGIS Instructions for... georeferencing an image without grid-lines

What are the six steps for georeferencing an image without grid-lines?

1. ....
2. ....
3. ....
4. ....
5. ....
6. ....

Create your own georeferenced map if you have one of a site you are working on by scanning it in and saving it as a .jpg then following the rest of your steps in *'10.4 My QGIS Instructions for... georeferencing a map image with grid-lines'*.

Congratulations! You have now georeferenced your own map image without grid-lines!

11. Habitat survey maps

Basics

You have returned from the field with a habitat survey map. You want to draw this in QGIS and produce a habitat map for your report. You also want to know the area of each habitat.

You will learn:

- More functions of QGIS desktop interface.
- How to create and edit vector data from field survey data.
- How to create points, lines and polygons.
- How to edit point, lines and polygons.
- How to troubleshoot in data creation.
- How to perform analysis on created data.
- More functions of QGIS print layout.
- How to produce a habitat map from created data.

11.1   Point, Line and Polygon

There are three types of shapefiles used by QGIS: Point, Line and Polygon.

We have worked with these different types in previous chapters, now we are going to learn how to create them by drawing or 'digitizing' them in QGIS.

Each shapefile type is suitable for digitizing different habitat categories (see guidance on UK Habitats Classification https://ukhab.org/).

In the left-hand column below is a list of habitats and on the right is a list of shapefile types. **Decide** which habitat category should be digitized as which shapefile, then **draw lines** between habitat category and shapefile type:

Hedge   Polygon

Tree   Line

Woodland   Point

Target note   Point

Pond   Line

River   Polygon

Did you come across any you were not sure of? Why do you think this is?

....

....

11.2 Steph's String Theory

Draw with a felt pen a dot onto the grey dot

You have created a **Point**.

Draw over the drawing of the zigzag using felt pen.

Redraw over this again, this time stopping to create

a dot at each corner. While you do this, imagine you

are taking a string and beginning at one end of the zigzag,

attaching the string with push pins at each corner.

You have created a **Line**.

When we draw in QGIS a line is created in the same way,

as if a string is attached to the mouse cursor.

In QGIS we click at each change in direction of a line to
create a 'vertex', which can later be edited.

Draw over the grey square with felt pen, stopping to

create a dot at each corner as before and imagine

you are taking a string and beginning at one

end, attaching the string using a push pin in each

corner, finishing the square with a fifth pin close to

the first pin.

You have created a **Polygon**.

When we draw in QGIS a polygon is created in the same way, as if a string is attached
to the mouse cursor. At each corner of the polygon a 'vertex' is created, which can
later be edited. When we draw in QGIS our final vertex of the polygon would be on
top of the first one.

Use the felt pen and imagine you have a string and push pins to 'digitize' the shapes
in the image below same way as the string theory exercise, applying understanding
from the matching exercise. (This is the same image we georeferenced in Chapter 10:
Georeferencing maps).

11.3   Basic Digitizing

We are going to digitize the polygons in QGIS in the same way as you have drawn on the map image above.

Create a new project titled **"Collymoon SSSI Habitats"**.

Add in the **raster** file that we georeferenced in the last chapter or use '**Collymoon Sketch grid georeferenced.tif**' provided.

Figure 11.1 Collymoon SSSI Georeferenced

If you are unsure about the accuracy of the georeferencing, make sure to add some transparency to the georeferenced image so you can cross reference the location as you digitize.

Alternatively, you can digitize by eye from the image (or a field map without georeferencing).

We are going to create a new vector layer.

In the **Layer** menu select '**Create Layer**' then '**New Shapefile Layer**' or use the '**New Shapefile Layer**' button in the left toolbar.

| Layer | Settings | Plugins | Vector | Raster | Database | Web | Mesh | Processing | Help |

Data Source Manager	Ctrl+L
Create Layer	▸
Add Layer	▸	Add Vector Layer...	Ctrl+Shift+V
Embed Layers and Groups...		Add Raster Layer...	Ctrl+Shift+R
Add from Layer Definition File...		Add Mesh Layer...
Georeferencer...		Add Delimited Text Layer...	Ctrl+Shift+T
Copy Style		Add PostGIS Layers...	Ctrl+Shift+D
Paste Style		Add SpatiaLite Layer...	Ctrl+Shift+L
Copy Layer		Add MSSQL Spatial Layer...
Paste Layer/Group		Add Oracle Spatial Layer...	Ctrl+Shift+O
Open Attribute Table	F6	Add SAP HANA Spatial Layer...
Filter Attribute Table	▸	Add/Edit Virtual Layer...
Toggle Editing		Add WMS/WMTS Layer...	Ctrl+Shift+W
Save Layer Edits		Add XYZ Layer...
Current Edits	▸	Add Vector Tiles Layer...
Save As...		Add WCS Layer...
Save As Layer Definition File...		Add WFS Layer...
Remove Layer/Group	Ctrl+D	Add ArcGIS REST Server Layer...
Duplicate Layer(s)		Add Vector Tile Layer...
Set Scale Visibility of Layer(s)		Add Point Cloud Layer...
Set CRS of Layer(s)	Ctrl+Shift+C
Set Project CRS from Layer	
Layer Properties...	
Filter...	Ctrl+F
Labeling	
Show in Overview	
Show All in Overview	
Hide All from Overview	

Figure 11.2 Layer menu <Create Layer> New shapefile Layer

In the pop-up box on the far right-hand side after '**File name**', use the '...' button to navigate to where you want to save the shapefile and type "**Collymoon Habitats Polygons**" to name your shapefile.

For '**Geometry type**' use the dropdown arrow and select '**Polygon**'.

Check the CRS is OSGB36/ British National Grid EPSG 27700.

Type "Habitat" in the '**Name**' box under the '**New Field**' heading, leave the 'Type' as 'Text data'.

Change 'Length' to '250'.

Click the 'Add to fields list' button.

What do you notice?

Click 'OK' at the bottom to create your shapefile.

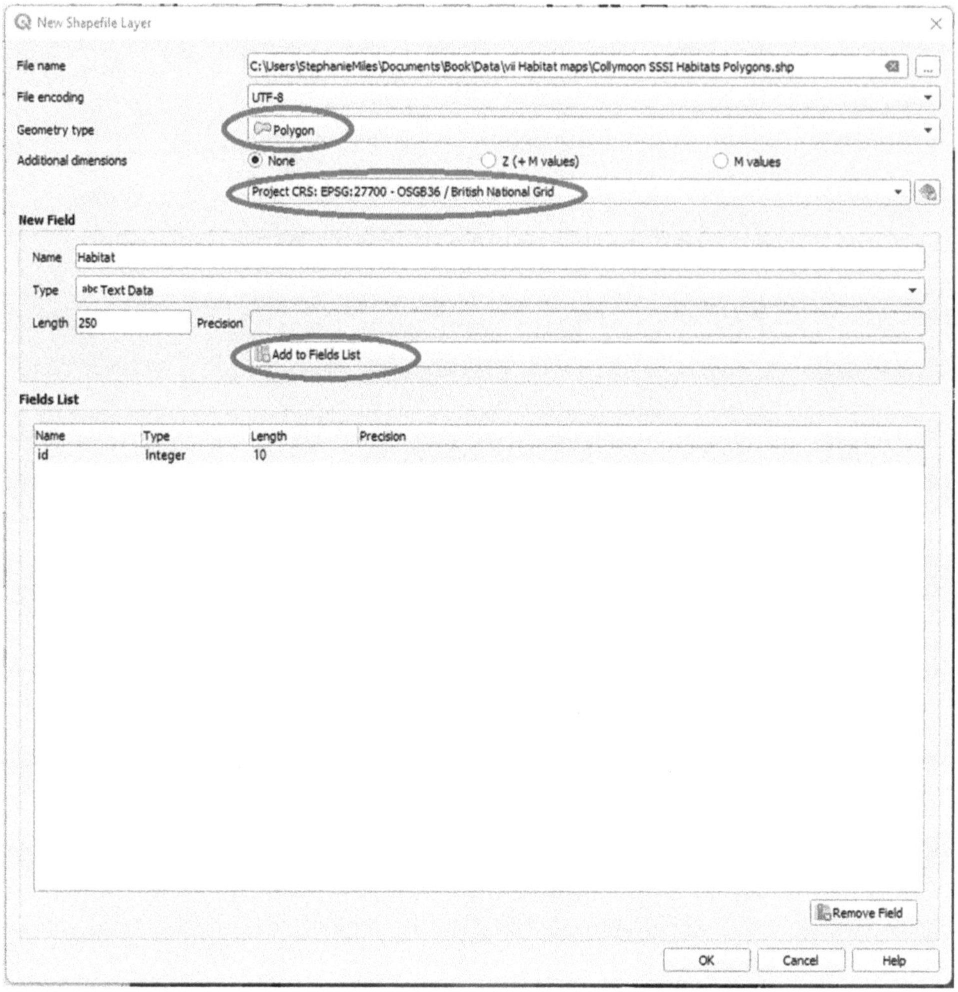

Figure 11.3 Create polygon

Congratulations! You have created your first polygon shapefile!

Once the layer has been created it will be added to the map.

Check the Attribute table for your new layer to see that the 'Habitats' field has been added.

Check the Project CRS and ensure it is set to British National Grid.

Repeat steps to create line and point shapefiles, selecting **'LineString'** and **'Point'** from the **'Geometry Type'** dropdown menu. Save as **"Collymoon SSSI Habitats Lines"** and **"Collymoon SSSI Habitats Points"**.

Digitizing

When digitizing shapes, you want to create the simplest shape possible to describe a feature, do not create false precision by adding many unnecessary points. This will only create more work for you in time spent digitizing and later on if you need to edit the shapes.

We are now going to digitize the shape of the heath habitat from the georeferenced image.

Click on **"Collymoon SSSI Habitats Polygons"** in the Layers panel then **click** the 'Toggle editing' button from the digitizing toolbar. This button starts and stops editing.

Figure 11.4 Toggle editing button

What do you notice?

....

Figure 11.5 Polygon digitizing toolbar

Click 'Add features button', this will put the cursor in digitizing mode.

Figure 11.6 Add features button

To digitize the 'Heath' as a polygon, **click** on the map area in the top corner of the area to create the first point of your new feature.

Working clockwise, **click** at each corner of the area for each additional point.

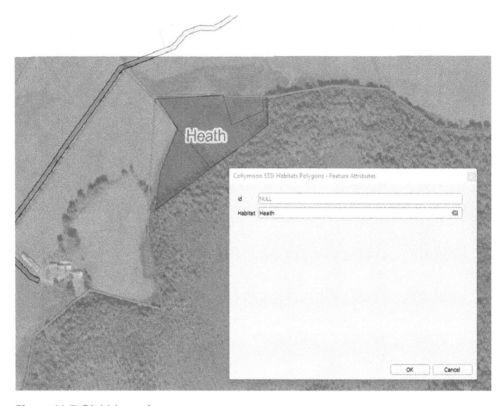

Figure 11.7 Digitizing polygon

Once you've added the final point, **right-click** to finish the shape.

The attribute window will appear to allow you to enter information for that feature.

Type "**Heath**" in the '**Habitats**' field.

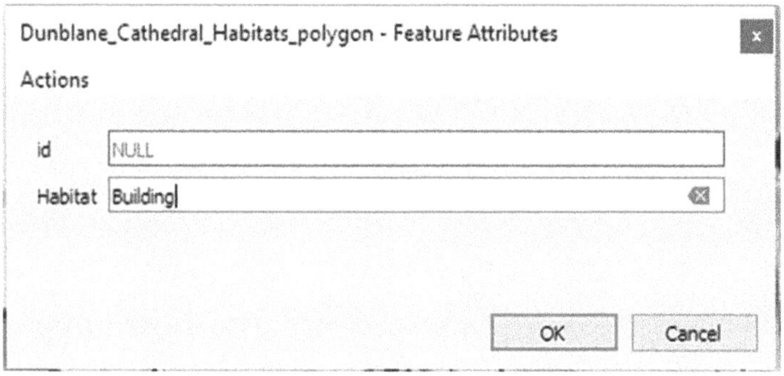

Figure 11.8 Digitizing pop-up

Click the **Save edits button** to save.

Figure 11.9 Save edits button

Click the Toggle editing button to stop editing, this will also prompt you to save changes.

If you want to start again, Click on "**Collymoon SSSI Habitats Polygons**" in the **Layers panel**.
Click the '**Select feature' button** from the Selection toolbar.

Figure 11.10 Selection toolbar

From the Digitizing toolbar **click** on the '**Delete feature**' button to delete your polygon representing the heath.

Figure 11.11 Delete features in the polygon Digitizing toolbar

Click on "**Collymoon SSSI Habitats Lines**" in the Layers panel then click the '**Toggle editing**' button from the digitizing toolbar.

Figure 11.12 Line digitizing toolbar

To digitize the Woodland ride as a **line**, click the **add line features button.**

Figure 11.13 Add line features button

Left-click on the map to start the line. Keep clicking for each additional point. **Right-click to finish.**

Type 'Ride' into the **Habitats field** in the pop-up box.

Save edits and untoggle editing.

Click on **"Collymoon SSSI Habitats Points"** in the Layers panel then **click** the **'Toggle editing'** button from the digitizing toolbar. This button starts and stops editing.

To digitize the nest as a **point**, use the **'add features' button.**

Figure 11.14 Point digitizing toolbar

Left-click to add the 'Nest' as a point. To keep this as a habitat rather than target note, type 'Tree' into the Habitats field.

Figure 11.15 Add point features button

You should now have something that looks like Figure 8.16.

Figure 11.16 Digitized habitats

11.4   My QGIS Instructions for... creating new shapefiles and basic digitizing

What are nine the steps for creating new shapefiles and basic digitizing?

1.   ....

2.   ....

3.   ....

4.   ....

5.   ....

6.   ....

7.   ....

8.   ....

9.   ....

11.5   Basic Habitat Map

We are going to make a habitats map using the habitat shapefiles you have created.

Add in the 'Collymoon_SSSI_boundary.shp'

In the left layers panel, click the box to **untick** the georeferenced image so the layer is no longer turned on. Ensure all the layers you want to show in the export are turned on.

In the layer properties turn on labels for each of your three habitats layers: points, lines, polygons and change the styles in the symbology and/or describe these in the legend.

Look back at '9.6 My QGIS Instructions for... creating a map using categorized styles and legend' to make a map layout and follow these to make a habitat map.

Ensure you add the OS copyright and the statement:

"Data derived from: Esri, Maxar, GeoEye, Earthstar Geographics, CNES/Airbus DS, USDA, USGS, AeroGRID, IGN, and the GIS User Community"

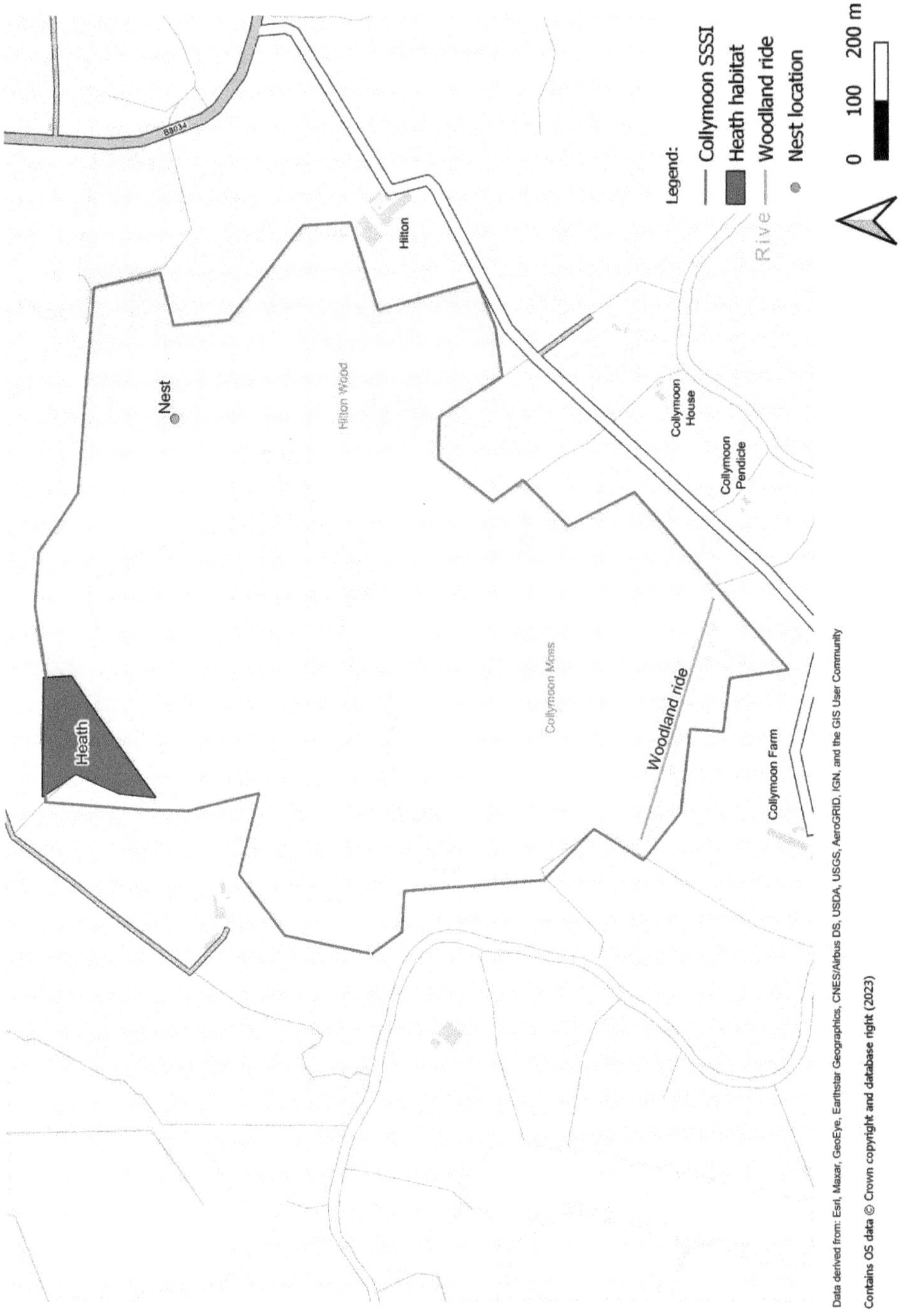
B8034
Nest
Hilton Wood
Hilton
Collymoon House
Collymoon Pendicle
Heath
Collymoon Moss
Woodland ride
Collymoon Farm
Rive
Legend:
Collymoon SSSI
Heath habitat
Woodland ride
Nest location
0 100 200 m
Data derived from: Esri, Maxar, GeoEye, Earthstar Geographics, CNES/Airbus DS, USDA, USGS, AeroGRID, IGN, and the GIS User Community
Contains OS data © Crown copyright and database right (2023)

Figure 11.17 Basic Habitats Map export

Well done! You have now made a basic habitat map!

Intermediate

We are now going to build on our knowledge from the Basics part of the chapter to edit our habitat map.

11.6 Basic Edits for Digitizing

We are going to digitize the main area within the boundary as a single polygon for plantation woodland and heath mosaic from the georeferenced image using the **"Collymoon_SSSI_boundary.shp"** and a tool called 'Snapping'.

Click the box to the left of the georeferenced image in the left layers panel to turn on.

Right-click in the Top toolbar and click to tick the box to the left of '**Snapping toolbar'**.

Figure 11.18 Snapping toolbar

Click the '**Enable snapping'** button to turn on snapping.

Click the second dropdown menu under the 3 square button and **click** to select 'Vertex'.

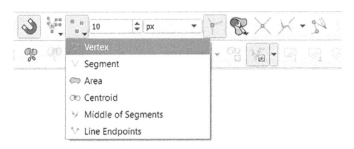

Figure 11.19 Snap to vertex

Set tolerance. This is the distance QGIS uses to search for the closest vertex. Select px (pixels) as the unit and 10 as the number. This will ensure that the snapping tolerance is the same irrespective of zoom.

Now snapping is on, we are going to use the **Vertex tool** to move the vertices of our heath polygon to match the boundary before we digitize any new features.

Click on "**Collymoon SSSI Habitats polygon**" in the **Layers panel.**

Click the '**Select feature' button** from the Selection toolbar.

Click on the 'Heath' polygon.

Click on the '**Toggle editing**' button from the Polygon Digitizing Toolbar.

Click on the **Vertex Tool** in the Polygon Digitizing Toolbar.

Click on the vertex you want to move, then **hover over** where you would like it to move to until the snapping 'magnetizes' to the vertex of the boundary, then **left-click**.

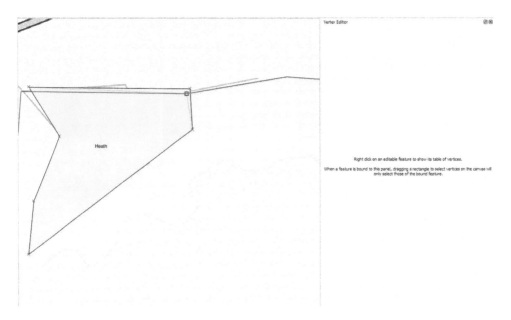

Figure 11.20 Moving vertex with snapping on

Optional practice: Repeat the vertex editing process to edit the boundary line and mosaic habitat to better fit the habitats boundary.

Draw the main plantation woodland and raised bog mosaic, going over the top of the woodland ride line but going around the edge of the heath polygon and fill in the 'Habitats' attribute of the new polygon. (In the *Advanced* section we will use new tools to split these habitats up.)

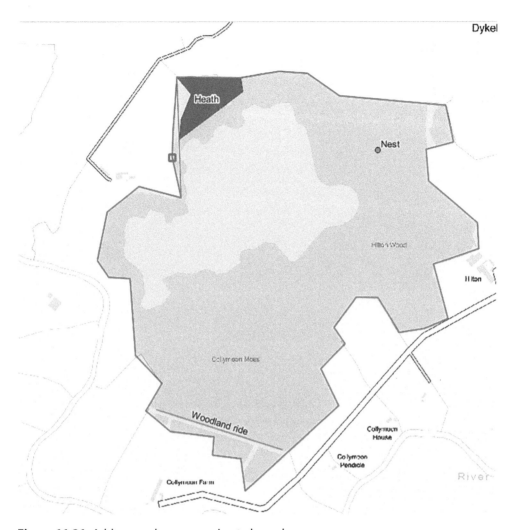

Figure 11.21 Add new polygon snapping to boundary

Describe what the snapping toolbar does:

....

....

....

Describe in your own words what the following buttons do:

....

....

....

....

....

Figure 11.22 Snapping toolbar naming exercise

11.7 Adding style files

Open the **Attribute table** for "**Collymoon SSSI Habitats polygon**", **Toggle editing** and **Add Field** to add a **Text(String)** field called '**p1code**' of length **12**. **Toggle off editing** and **Save edits**. Check attribute table for your new field.

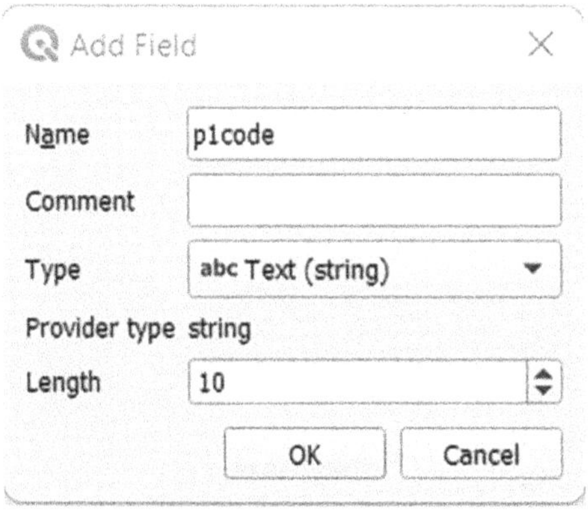

Figure 11.23 Add field

In **Layer Properties, Symbology** tab at the bottom left of the window **click** the '**Style**' menu.

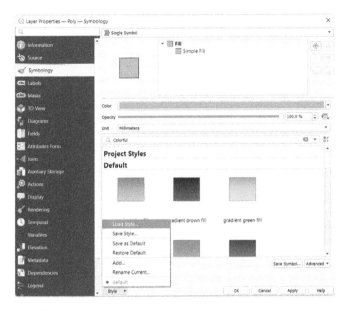

Figure 11.24 Load Style

Use the '...' button to navigate to the polygon style file **"P1 Habitat Survey Toolkit QGIS QML Polygon Style File.qml"**.

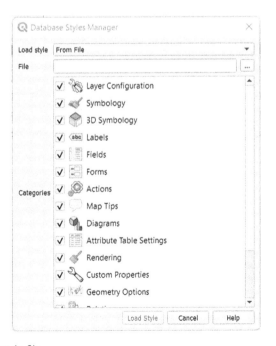

Figure 11.25 Load style file

Phase 1 habitats styles

These can be downloaded from the Field Studies Council, from this webpage at time of writing: https://www.fscbiodiversity.uk/qgis-mapping-styles-uk-habitats

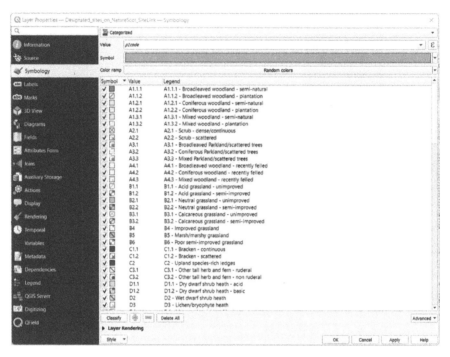

Figure 11.26 Phase 1 habitats styles

Open the **Attribute table** for "**Collymoon SSSI Habitats polygon**", **Toggle editing** and **Type** in the **Phase 1 habitat codes** to the '**p1code**' field. If you do not know the code, you can look these up in the loaded symbology. The habitat code needs to be entered identically to as it is written in the style symbology e.g. "A1.1.1" will categorize correctly but "A.1.1.1" and "A1.1" will not.

For Heath use 'D6' and for the mosaic use 'J5', for Wet heath/acid grassland and Other habitat respectively.

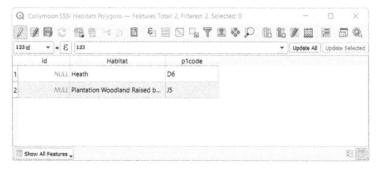

Figure 11.27 Habitats Attribute table with p1codes

The style will then be available for all future QGIS Projects.

What is a style file and how do you load one?

....

....

....

For the purposes of the Intermediate Habitat Map export, change J5 in the symbology to transparent.

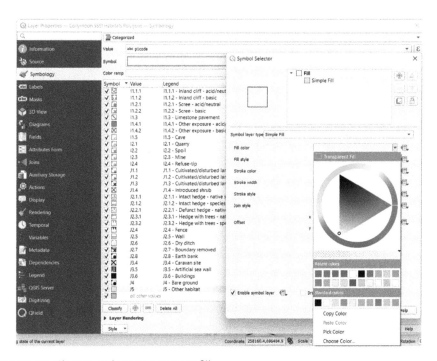

Figure 11.28 Change style to transparent fill

Repeat the process for **"Collymoon SSSI Habitats lines"** and **"Collymoon SSSI Habitats points"**. **Add p1code fields to each attribute table. Load style files** (.qml): "P1 Habitat Survey Toolkit QGIS QML Line Style File with unclassified.qml" and "P1 Habitat Survey Toolkit QGIS QML Target Note Style File.qml".

UK Habitats

UK Habitat Classification has recently been published as a replacement for the Phase 1 habitat survey. It specifies what size of habitat should be mapped as polygon, line and point. For more information consult http://ecountability.co.uk/ukhabworkinggroup-ukhab/

11.8 Calculating Areas

Now we are going to use the field calculator to calculate the areas of the habitats. Open the attribute table of "**Collymoon SSSI Habitats Polygons**".

Click the **field calculator** button.

Figure 11.29 Field calculator button

Select **Create a new field**, type an output field name '**Area**'.
Select **Whole number (integer)** from the Output field type.
Under functions, select **Geometry**.
Double-click '$area'. Click OK.

Figure 11.30 Field calculator

A new column will be added to the attribute table with the area calculated.

What units has the area been calculated in and why?

....

We want the area in hectares, how do we get the field calculator to calculate this?

....

Compare the use of the measuring tool to the field calculator.

....

....

....

....

....

Which do you prefer and why?

....

....

....

....

11.9   My QGIS Instructions for... calculating polygon areas

What are the six steps for calculating polygon areas?

1.   ....

2.   ....

3.   ....

4.   ....

5.   ....

6.   ....

11.10   Intermediate Habitat Map

We are going to make a habitat map using the habitat shapefiles you have created.

Look back at '**9.6 My QGIS Instructions for... creating a map using categorized styles and legend**' to make a map layout and follow these to make a habitat map. This time use the UK Habitats style file to categorize the polygon styles. Use the 'Insert Attribute table' button to add in the attribute table to the legend, to display the area of each polygon.

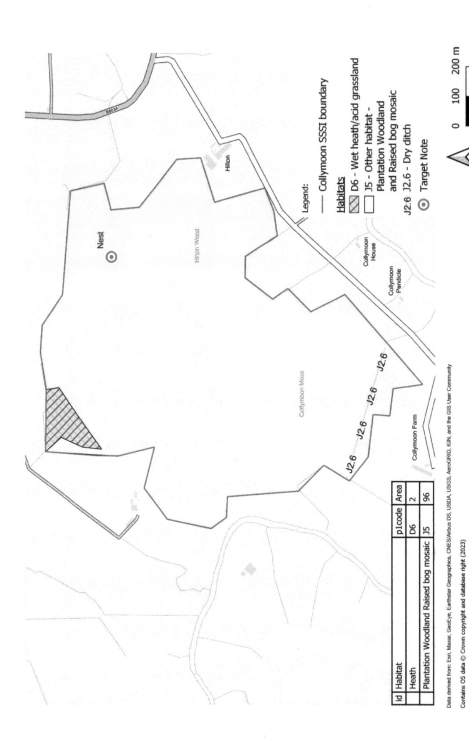

Figure 11.31 Intermediate Habitats Map export

Ensure you add the OS copyright and the statement:

Data derived from: Esri, Maxar, GeoEye, Earthstar Geographics, CNES/Airbus DS, USDA, USGS, AeroGRID, IGN, and the GIS User Community

Superb! You have now made an improved habitat map!

Using your georeferenced map from Chapter 10 and following '10.4: My QGIS Instructions for... creating new shapefiles and basic digitizing' create your own habitats map of your site using '9.6 My QGIS Instructions for... creating a map using categorized styles and legend' to make a map layout and follow these to make a habitat map.

Congratulations! You have now made a habitat map of your site!

Advanced

We are now going to build on our knowledge from the *Intermediate* part of the chapter to use some of QGIS's advanced editing features to improve our habitat map.

11.11   Advanced edits for digitizing

We are going to use the advanced digitizing tools to create more polygons.

Right-click the top toolbar menu and turn on the '**Advanced Digitizing**' toolbar.

Figure 11.32 Advanced Digitizing toolbar

Select the plantation woodland and raised bog mosaic polygon. Use the **Add ring** button to add a hole in polygon corresponding with the change in habitat between the plantation and bog habitat.

Figure 11.33 Add ring

Digitize only within the existing polygon.

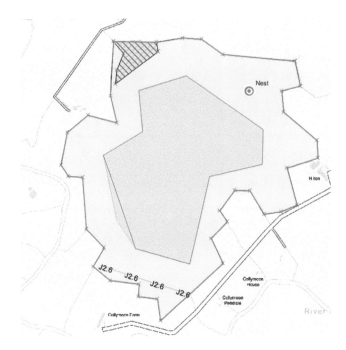

Figure 11.34 Add ring to polygon

Use the **Vertex tool** to add more vertexes and move existing ones until you get the desired shape.

Figure 11.35 Edit vertexes in add ring

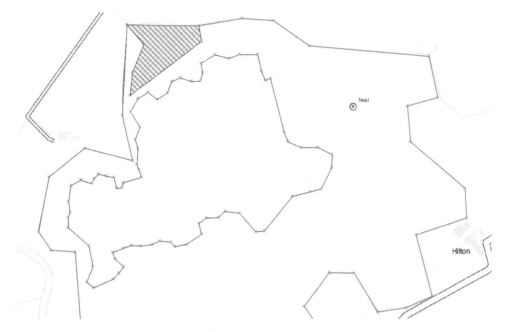

Figure 11.36 Complete vertexes in add ring

If you need to delete the ring and start again, select the '**Delete ring**' button, **click** inside ring and delete.

Figure 11.37 Delete ring

Select the plantation polygon, now containing a hole corresponding to the bog habitat. **Add** the central raised bog area inside the plantation by using the '**Fill Ring**' button.

Figure 11.38 Fill ring

Update the attributes table and symbology to reflect the changes.

We want to add a gap for the woodland ride in the Habitats polygon shapefile.

Select the plantation polygon. Use '**Split Features**' to draw a line across the plantation polygon where you want to split. Start outside the polygon, cross it and right-click outside. Two polygons will now be created.

Figure 11.39 Split features tool

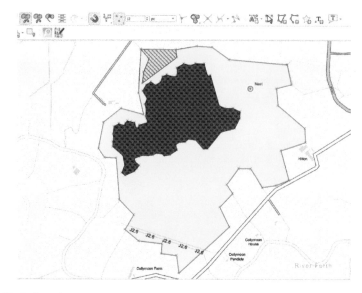

Figure 11.40 Split polygon

Open the attribute table, what has happened?

....

Split the south polygon again to create a polygon to cover the woodland ride area.

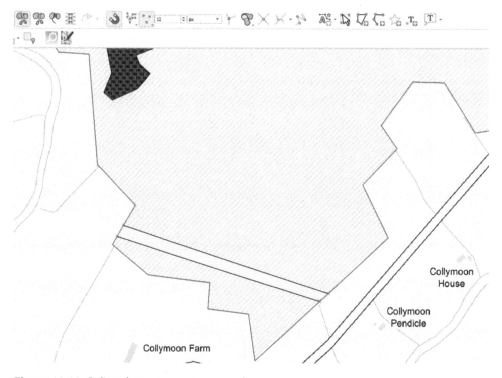

Figure 11.41 Split polygon to create new polygon

Use **Reshape features** and to do finer edits to the shapefiles.

Figure 11.42 Reshape features

Select the feature. Use the '**Reshape Features**' tool to add to the polygon: Click inside polygon, cross the boundary, add a series of new vertices outside the polygon and then finish by right-clicking inside the polygon.

To remove part of the polygon: do the reverse. Start outside, cross the boundary, add any vertices inside the polygon and right-click outside the polygon.

Calculate areas for your habitat polygons as we did in the previous exercise. This time, you will need to select 'Update existing field' instead of 'Create a new field'. Make sure 'Only update 1 selected features' is unticked if you have a feature selected in the map or attribute table.

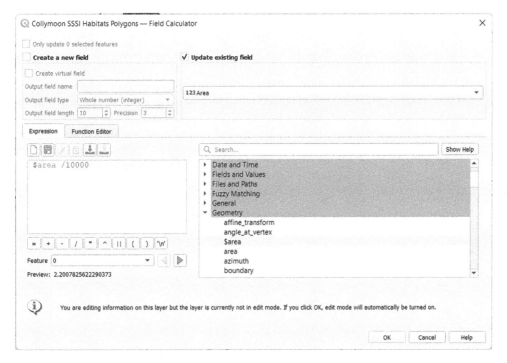

Figure 11.43 Field calculator update existing field

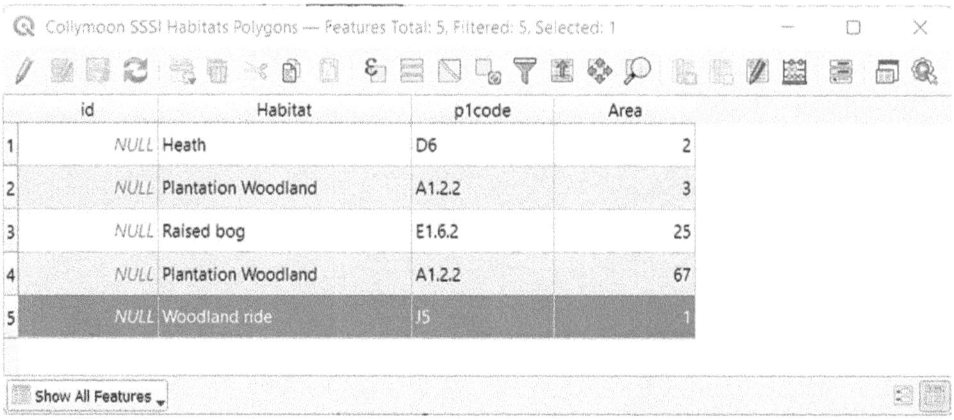

Figure 11.44 Updated attribute table

We want to calculate the area of the whole site, but the boundary is currently a line.

Create a new polygon shapefile for the boundary, digitize it using the '**Snapping toolbar**' and **calculate** the area in hectares. Check this matches the sum of the habitats.

11.12 Advanced Habitat Map

Look back at '**9.6 My QGIS Instructions for... creating a map using categorized styles and legend**' to make a map layout and follow these to make a habitat map. This time use the Phase 1 Habitats style file to categorize the polygon styles. Use the insert Attribute table button to add in the attribute table to the legend, to display the area of each polygon.

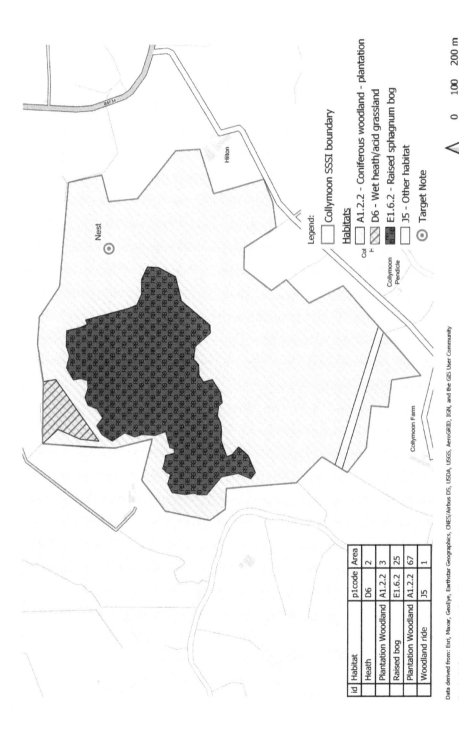

id	Habitat	p1code	Area
	Heath	D6	2
	Plantation Woodland	A1.2.2	3
	Raised bog	E1.6.2	25
	Plantation Woodland	A1.2.2	67
	Woodland ride	J5	1

Data derived from: Esri, Maxar, GeoEye, Earthstar Geographics, CNES/Airbus DS, USDA, USGS, AeroGRID, IGN, and the GIS User Community

Contains OS data © Crown copyright and database right (2023)

QGIS mapping styles for UK habitats downloaded from: https://www.fscbiodiversity.uk/qgis-mapping-styles-uk-habitats

Figure 11.45 Advanced Habitats Map export

Well done! You have now made an advanced habitat map!

Using your georeferenced map of your own site from Chapter 10 and following '10.4: My QGIS Instructions for... creating new shapefiles and basic digitizing' create your own habitats map of your site using '9.6 My QGIS Instructions for... creating a map using categorized styles and legend' to make a map layout and follow these to make a habitats map.

Ensure you add the OS copyright and the statements:

Data derived from: Esri, Maxar, GeoEye, Earthstar Geographics, CNES/Airbus DS, USDA, USGS, AeroGRID, IGN, and the GIS User Community

AND

QGIS mapping styles for UK habitats downloaded from: https://www. fscbiodiversity.uk/qgis-mapping-styles-uk-habitats

Excellent! You have now made an advanced habitat map of your site!

And congratulations you have completed the workbook!

Part III: Answers to Exercises

5. Basic maps

5.1 The QGIS interface

What do you notice?
Name of tool appears over button

What is the button below, what does it do and where can it be found?
New Project button, opens a new QGIS Project. It is in the top left-hand corner of the QGIS opening screen.

What are the seven parts of the interface on the QGIS Project screen?

1. Top menu
2. Top toolbars
3. Shortcut panel
4. Browser panel
5. Layers panel
6. Map area
7. Bottom info bar

5.2 Rasters and Vectors

Write down differences between each layer:

Vector:

1. *It is a line*
2. *It is editable*
3. *It has an Attribute table*

Raster:

1. *It is a picture*
2. *It is not editable*
3. *It has no Attribute table*

5.3 Shapefiles and GeoTIFFs

In your own words, define Vector and Raster data:

A Vector is:
A collection of files that make up a editable spatial layer shape when viewed in a GIS system.

A Raster is:
A collection of files that make up a non-editable spatial layer image when viewed in a GIS system.

5.4   Map navigation

Pan: lets you move around the map

Zoom in: Click and drag to draw a box to zoom in

Zoom out: Click and drag to draw a box to zoom out from

Zoom to extent of all layers

Previous extent

Next extent

5.5   Coordinate Reference Systems

What do you notice about the basemap?
Map has been skewed.

Why has this happened?
We have changed the coordinate reference system from OSGB 36 which is in metres to WGS 84 which is in degrees.

Why did we need to select a transformation?
When we changed the CRS we changed the measurement system from metres to degrees, QGIS calculates how to display this for us using transformations.

5.6   Map-making basics

What has happened to the basemap image and why?
The basemap image is pixellated, this is because when we zoomed to the boundary line layer we zoomed in further than the scale of the raster basemap.

The layout screen:

1. *Left toolbar – contains the tools for editing the canvas.*
2. *Canvas – the drawing area for the map export.*
3. *Items panel – shows the item order in the canvas.*
4. *Right panel – contains tabs for editing layout, Item properties and guides.*

On the left-hand side is a toolbar, as you use these buttons in the following exercise, write down what each tool does next to the image*:

Pan layout
Zoom
Select/Move item
Move item content
Edit nodes item
Add map
Add 3D map
Add picture
Add label
Add legend
Add scalebar
Add north arrow

Add shape
Add marker
Add arrow
Add node item
Add HTML
Add Attribute table
Add fixed table

*order and wording may be slightly different in different versions of QGIS

Circle and label the buttons below:

Save project
New layout
Duplicate layout
Layout manager
Add items from template
Save as template
Add pages
Print layout
Export as image
Export as svg
Export as pdf
Undo
Redo
Atlas toolbar (greyed out)
Zoom toolbar
Actions toolbar

5.7   My Instructions for... exporting a basic map using QGIS

List the six steps you used to produce your basic map:

1. *Open New layout*
2. *Draw in new map*
3. *Edit scale*
4. *Add scalebar*
5. *Add copyright statement*
6. *Export as image*

5.10   My Instructions for... downloading and loading basemaps

What are the seven steps for downloading and loading basemaps into QGIS?

1. *Go to OS Open Data website*
2. *Find the type of data you want to download*
3. *Select which area*
4. *Select data type*
5. *Click download and save file*
6. *Unzip file*
7. *Use Add layer>Add raster to add into QGIS*

6. Survey maps

6.1 Connecting to online basemaps

What are the differences between using a downloaded Raster tile (i.e. the Ordinance Survey map) and XYZ tiles (i.e. the OpenStreetMap)?
Different styling of imagery, online mapping may load a bit slower on zoom in and out, different scales of imagery layers are loaded automatically rather than a single image

What kind of data is OpenStreetMap and how can you tell?
Raster, due to the raster icon and no Attribute table

6.3 My Instructions for... connecting to online basemaps

Describe in your own words how to connect to online basemaps.

Find XYZ tiles in Browser panel, use dropdown arrow to show OpenStreetMap and click and drag this into the Layers panel.

6.7 My Instructions for... exporting a survey map using QGIS

List the eight steps you used to produce your survey map:

1. *Open New layout*
2. *Draw in new map*
3. *Edit scale*
4. *Add scalebar*
5. *Add copyright statement*
6. *Add north arrow*
7. *Add grid-lines*
8. *Export as image*

6.8 My Instructions for... basemap copyright

Why is it important to add a copyright statement to every map you produce?
To show who the copyright of the maps and data belongs to and to comply with the conditions of the license.

What is the copyright statement for Ordnance Survey maps and data?
"Contains OS data © Crown copyright and database right (year)"

What is the copyright statement for OpenStreetMap?
"(c) OpenStreetMap contributors"

What is the copyright statement for ESRI Imagery?
"Source: Esri, Maxar, GeoEye, Earthstar Geographics, CNES/Airbus DS, USDA, USGS, AeroGRID, IGN, and the GIS User Community"

7. Designated sites map

7.1 Setting up a Project: Setting the Coordinate System

What is a coordinate reference system (CRS)?
A coordinate reference system is the way the GIS system understands where the layer is in the world. There are different systems for each country and world coordinate systems.

Why is it important to set a CRS when setting up a project?
To tell QGIS how to read the files so it knows where to display them.

7.2 Setting up a Project: Importing vector and raster data

What is the difference between a shapefile and a project?
A shapefile is a group of files you can edit and display in a project. A project is where you can view layers both shapefiles (vector) and images (raster) aligned to where they are geographically.

What is the difference between a map and a project?
A map is the static image you create in the layout window and export for printing. A project is where you make the map and store the links to the layers in.

7.3 My QGIS Instructions for... setting up a Project

What are the four steps for setting up a project?

1. *Open new project*
2. *Set CRS*
3. *Import shapefiles and basemap*
4. *Save Project*

7.4 Using QGIS tools

What is the SSSI name?
Arthur's Seat Volcano

What do you notice when you use the identify features tool on the raster layer?
Information about the name of the layer colour band of the image is displayed no attribute information.

Find and use the measure tool to measure the distance between the site boundary and SSSIs:
Arthur's Seat Volcano SSSI, 0m from site boundary
Duddingston Loch SSSI 2,300m from site boundary
Wester Craiglochart Hill SSSI 3,800m from site boundary

Use the measure tool to measure the area of each SSSI:
Arthur's Seat Volcano SSSI 5ha + 9ha + 220 ha

Duddingston Loch SSSI 25 ha
Wester Craiglochart Hill SSSI 4ha

What happens?
SSSI is highlighted in yellow

What information do you see in the table for your chosen SSSI?
Name, Status, Area in hectares, Type etc.

7.6 QGIS tools test

Describe what the following buttons do in your own words:
Select tool: Highlights a feature from a shapefile which can then be looked at in the Attribute table
Identify tool: Brings up layer information
Text annotation tool: Lets you add a label to the map
Measure line tool: Allows you to measure between multiple points

7.7 My QGIS Instructions for... measuring and recording distances

Describe how you measure distances between shapefiles and how to record them in the Attribute table.

1. *Select 'Measure line' tool.*
2. *Click on the edge of one shapefile boundary then measure over to the other and click. Sense check this distance.*
3. *Open Attribute table and toggle editing.*
4. *Add a field called 'Distance' accepting the defaults.*
5. *Click in the cell and type in the distance measured.*
6. *Click to save edits and toggle off editing.*

7.8 Styling lines and polygons

Why is it useful to be able to change the appearance of vector layers in a map?
Vector layers often overlap each other and being able to differentiate between them and make them partially translucent is useful.

7.9 Labelling

Open the 'Value' dropdown menu, what is this value and where does it come from?
The value menu has the 'Name' field selected from the attributes table of the layer.

Why is this useful?
We can automatically label the layer by any field in the attributes table.

Hoover the mouse cursor over a label, what do you notice?
The label is within a red box to show you it is selectable.

Compare the use of annotation and labelling, which do you prefer and why?
Annotation allows you to directly create text boxes on the map.

Labelling allows you to use fields in the attribute table to label the layer features.

Labelling looks better and is easier for labelling multiple features, annotations are simpler and allow labelling of individual features rather than every one.

7.11 My QGIS Instructions for... creating a map using styles and legend

Describe in your own words how to change the symbology of a vector file
Right-click on vector file in layer panel, left-click on 'Properties' then 'Symbology' tab. Use the options to change colours, lines and shapes of symbols.

List the eight steps you used to produce your SSSI map:

1. *Open New layout*
2. *Draw in new map*
3. *Edit scale*
4. *Add scalebar*
5. *Add copyright statements*
6. *Add north arrow*
7. *Add legend*
8. *Export as image*

Describe in your own words how to add and alter a legend in the layout
Use 'Add image', click and drag to draw box. In the right panel deselect 'Auto update' and add in or take out layer legends using + and – buttons and reorder using up and down buttons.

7.12 Designated Sites map

What do you notice about the layers?
There is a white background which hides the other layers in the map.

What do you notice about the layer in the map?
Changing the transparency/opacity means we can now see the layers underneath.

Do you notice a difference between the appearance of these layers and the layers we connected to for Protected Areas in Scotland?
The Protected Areas in Scotland had a white background that obscured the other layers but the English/Welsh ones do not.

Why do you think this is?
The Natural England/Natural Resources Wales layers are Web Feature Services so are vector layers, whereas the Nature Scot layers are Web Map Services so are raster layers.

7.13 My Instructions for... downloading vector data from government sources, connecting to government web map services and government data copyright

What are the four steps for downloading vector data from a government website?

1. *Visit NS/NE/DMW/ODNI website*
2. *Search for required dataset*
3. *Download data*
4. *Look up copyright statement*

What are the six steps for setting up a new connection to a web map service from a government website?

1. *Visit NS/NE/DMW/ODNI website*
2. *Search for required dataset*
3. *Look up copyright statement*
4. *Copy url of WMS/WFS*
5. *Create new connection in Data Source Manager*
6. *Paste url into connection window*

What is the copyright statement for Nature Scot data?
"© SNH, Contains Ordnance Survey data © Crown copyright and database right (2022)."

What is the copyright statement for Natural England data?
"© Natural England copyright. Contains Ordnance Survey data © Crown copyright and database right [year]."

What is the copyright statement for Welsh Government/Natural Resources Wales data?
"Contains Natural Resources Wales information © Natural Resources Wales and Database Right. All rights Reserved. Contains Ordnance Survey Data. Ordnance Survey Licence number 100019741. Crown Copyright and Database Right."

What is the copyright statement for Environmental Protection Agency Northern Ireland data?
"©NIEA, 2019 [dataset name] is licensed under the Open Government Licence: http://www.nationalarchives.gov.uk/doc/open-government-licence/version/3/"

8. Desk-study maps

Creating point data: Importing from a spreadsheet

What information is contained in the first row?
Column headings

What data is contained in the OSGR column?
Grid reference

What is the difference between the data in the OSGR column and the two preceding columns?
Grid reference is letters and numbers, next two are Eastings and Northings which are number only.

8.2 *My QGIS Instructions for... installing plugins*

What are the four steps for installing a plugin?

1. *Go to Plugins then Manage and Install Plugins*
2. *Select All tab*
3. *Search for required plugin*
4. *Click install and wait for install, restart QGIS if new toolbar/menu does not appear*

Why do we need to convert to a shapefile?
To be able to work with the file in QGIS, to perform edits and change symbology.

8.3 My QGIS Instructions for... importing Ordnance Survey National Grid Coordinates

What are the six steps for importing a spreadsheet using the TomBio Toolbar Biological records tool?

1. *Open the Biological records tool*
2. *Browse to the csv file you want to import using '...' button*
3. *Select the grid reference column from the "OS Grid Ref" dropdown menu*
4. *Select "Records as points" or use this dropdown menu to map as grid square with scale of your choice*
5. *Click create map layer*
6. *Save as shapefile*

8.5 Desk-study map from a spreadsheet

What is the copyright statement for the basemap?
"(c) OpenStreetMap contributors"

Creating point data: Importing data from NBN

8.6 Downloading data from NBN

What does the data show?
Coordinates at different accuracy/scales

Describe the output on the map
Polygons of 10km square

Compare the Attribute table for the shapefile imported by mapping "Records as points" with Attribute table of the shapefile imported by mapping "Atlas squares" and the original CSV file, what do you notice? Why do you think this is?
Less fields in the Atlas squares as data has been amalgamated across grid squares. Original data preserved in the records as points import.

Which method would be useful for which circumstances?

1. **Records as points import:** *when you have high accuracy data to display on a small scale and you want to import all csv columns.*

2. **Atlas squares import:** *when you have data with lower accuracy or want to show it on a large scale and want to summarize data by abundance, e.g. creating an atlas.*

Which do you think would be most applicable to your work and why?

Records as points – I have accurate records to display at a large scale.

Atlas – I have lots of data over a large area I want to summarize.

8.7 My QGIS Instructions for... downloading data from NBN

What are the six steps for downloading data from NBN?

1. *Go to NBN*
2. *Search for data and check licence or request access*
3. *Download data*
4. *Unzip file*
5. *Add into QGIS using TomBio Toolbar>Biological records tool*
6. *Save features as shapefile*

Connecting to NBN data

What scale are the Atlas squares and why?

10km square, because that is the default of data accuracy on the NBN to for public availability.

What do you notice about the layer when it appears in the Layers panel?

It is a raster.

Why do you think this is?

Because WMS provide rasters.

What would happen if we saved the features in this layer?

We would have a raster dataset.

8.8 My QGIS Instructions for... connecting to data from the NBN

What are the seven steps for connecting to data from the NBN?

1. *Go to NBN*
2. *Search for data and check licence or request access*
3. *Open NBN Atlas Tool*
4. *Use Filter tab to search for dataset*
5. *Use NBN tab to map dataset*
6. *Use NBN tab to download dataset*
7. *Add dataset into QGIS using TomBio Toolbar>Biological records tool*

Which of the methods for downloading and importing data from the NBN do you prefer and why?

Connecting directing to the NBN via the NBN Atlas Tool and importing via the Biological records tool lets you connect directly to the data. However, need an internet connection and to check the data on the website to see if you can get better access. N.B. Sometimes this direct access to the NBN may not work, for instance because the site is under maintenance.

8.9   Desk-study map from NBN data

What is the copyright statement and where can it be found?
"Records provided by Natural Resources Wales, accessed through NBN Atlas website." Found in citation.csv downloaded with data.

9.   Protected species survey map

Creating point data: importing from a GPS

9.1   Importing GPX files

Why did we select Projection ESPG:4326 – WG84 instead of EPSG:27700 - OSGB36 / British National Grid?
GPS stands for Global Positioning Satellite. Because it uses the position on the globe rather than within the UK the coordinates are recorded in a world coordinate reference system, this is usually ESPG:4326 – WG84.

9.2   My QGIS Instructions for... importing from a GPS

What are the seven steps for importing from a GPS?

1. *Upload GPS files to computer, convert GPS to GPX*
2. *Open Data Source Manager and select GPS tab*
3. *Browse to file*
4. *Select waypoints to import points or tracks to import a recorded track you have walked*
5. *Export as shapefile*
6. *Create spatial index*

9.3   Troubleshooting with GPS

Why might points taken from a GPS not plot on the map exactly where you expect?
They might be at a lower accuracy/only recorded to lower accuracy. A cloudy day or tree cover could have obscured GPS satellite triangulation. Wrong CRS.

Describe the ways you might change where a point plots on the map
Increase accuracy of recording. Resurvey on a cloud free day. Set GPS to BNG. Manually edit points using toggle editing (more on this later).

9.4   Categorized symbology

What information is stored in the 'Description' column?
Notes on what was found at each point.

How can we simplify this information in order to map it?
Create another column and summarize description into categories.

Why is it useful to categorize target notes by colour and to label them on the map?
To be able to reference your target note table and map in your report.

9.6   My QGIS Instructions for... creating a map using categorized styles and legend

Describe in your own words how to use categorized symbology to style a vector file.
Create a category column in your Attribute table to summarize each type of data you want to show on the map. In the symbology tab in the layer properties change 'single symbol' to 'categorized symbol' and change colours as desired.

List the eight steps you used to produce your Protected Species map:

1.   *Open New layout*
2.   *Draw in new map*
3.   *Edit scale*
4.   *Add scalebar*
5.   *Add copyright statements*
6.   *Add north arrow*
7.   *Add legend*
8.   *Export as image*

Describe in your own words how to add and alter a legend in the Layout with categorized vectors
Use 'Add legend' button to draw box for legend. Use the + and − buttons and up and down buttons to create and order the layer. Use fonts beneath to edit font.

10.   Georeferencing maps

10.1   Georeferencing using OS grid-lines

As you add each point what do you notice?
The points appear in the table below.

Why do the images not exactly match up?
The images are not the same.

The roads in the OS map are wider than the satellite.

The maps use different native CRS OS map is in BNG and satellite is World Mercator.

Open the folder where you saved the georeferenced image. What do you notice?
A GeoTIFF has been created.

10.2   Georeferencing without OS grid-lines

As you add each point what do you notice?
The points appear in the table below.

How do we check if our georeferencing is accurate?
Compare accuracy of fit to basemap by changing transparency of raster image.

10.3 My QGIS Instructions for... georeferencing an image with grid-lines

What are the six steps for georeferencing an image with grid-lines?

1. *Scan and save image as JPG.*
2. *Open the JPG in the Georeferencer window.*
3. *Add control points by using grid-line intersections and typing in coordinates from image.*
4. *Go to Transformation settings>Polynomial 1. Set target CRS.*
5. *Tick the 'Load in QGIS' box when done. Click the green arrow.*
6. *Check accuracy of georeferencing and repeat georeferencing if necessary.*

10.4 My QGIS Instructions for... georeferencing an image without grid-lines

What are the six steps for georeferencing an image without grid-lines?

1. *Scan and save image as JPG.*
2. *Open the JPG in the Georeferencer window.*
3. *Add control points by clicking in the image then using from map canvas" to click through to basemap and selecting equivalent point.*
4. *Go to Transformation settings>Thin Plate Spline. Set target CRS.*
5. *Tick the "Load in QGIS" box when done. Click the green arrow.*
6. *Check accuracy of georeferencing and repeat georeferencing if necessary.*

11. Habitat Survey maps

Basics

11.1 Point, Line and Polygon

On the left-hand side below is a list of habitats and on the right is a list of shapefile types. Decide which habitat type should be digitized as which shapefile, then draw lines between habitat type and shapefile:

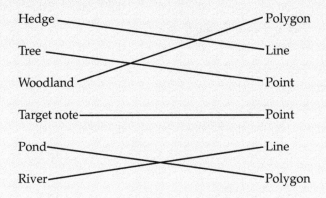

Did you come across any you were not sure of? Why do you think this is?
Polygon versus line versus point depends on the size of the feature and the scale at which it is to be mapped.

11.3 Basic Digitizing

What do you notice?
When you click Add to Fields List button, the new field appears in the New Fields List box.

What do you notice?
When you click the Toggle editing button the rest of the digitizing toolbar goes from greyed out to full colour meaning it is activated for use.

11.4 My QGIS Instructions for... creating new shapefiles and basic digitizing

What are nine the steps for creating new shapefiles and basic digitizing?

1. *Select 'New shapefile layer' from top toolbar.*
2. *Select folder to save in and type file name.*
3. *Select Geometry type.*
4. *Check CRS.*
5. *Click on layer in layer panel.*
6. *Click 'Toggle editing' button.*
7. *Click 'Add features' button.*
8. *Draw polygon/line/point.*
9. *Toggle off editing and save edits.*

Intermediate

11.6 Basic Edits for Digitizing

Describe what the snapping toolbar does
'Magnetizes' the mouse cursor to the nearest vertex/intersection/line when options are selected to the accuracy chosen in m/pixels etc.

Describe in your own words what the following buttons do:
Toggle editing: Turns on edits for the selected layer.

Add new feature: Click to draw new polygon.

Add new line: Click to draw new line.

Add new point: Click to draw new point.

Vertex tool: lets you select, move and delete individual vertexes of your drawing.

11.7 Adding Style files

What is a style file and how do you load one?
A style file is a premade symbology file that can be loaded into QGIS.

It can be loaded using the style menu at the bottom of the symbology tab in layer properties.

Tip: Check the 'Value' menu in the 'Categorized Symbology' window is linking to the correct field.

11.8   Calculating Areas

What units has the area been calculated in and why?
Square metres, as these are the units of the map

We want the area in hectares, how do we get the field calculator to calculate this?
$area/10000

Compare the use of the Measuring tool to the Field calculator
The field calculator is a much more accurate way to calculate areas.

The measuring tool is quick and easy but less accurate.

Which do you prefer and why?
Field calculator for automatically calculating areas

Measuring tool for quickly measuring distances

11.9   My QGIS Instructions for... calculating polygon areas

What are the six steps for calculating polygon areas?

1. *Open Attribute table and toggle editing.*
2. *Use field calculator to add new field called 'Area'.*
3. *Select 'Whole number (integer)'.*
4. *Double-click $area from the Operators menu or type in box and click 'OK'.*
5. *Sense check area calculation and convert to hectares if required.*
6. *Save edits and toggle off editing.*

Advanced

11.11   Advanced edits for digitizing

Open the attribute table, what has happened?
The newly split polygon is now two rows in the attribute table with the same attributes as the original.

Part IV: Workflows

Instructions for how to perform each task, to make each map and perform each analysis as described in the main text.

12.1 Basic maps

Step 1: OS Download
Go to the OS Open Data downloads page and download data for the area you are working in by Ordnance Survey grid square. To the most detailed basemap to show at 1:20,000–1:30,000 scale is listed on the website as 'OS Open Map – Local'. To find your area, use the 'Set a custom areas' and select map tile(s). Select 'GeoTIFF Full Colour'. Download.

Step 2: Unzip file
Unzip the folder you have downloaded.

Step 3: Open New Project
Open QGIS Desktop. Click 'New Project' button in the top toolbar.

Step 4: Save Project
In the top toolbar choose 'Save Project As', navigate to where you want to save the project and type a file name. Click 'Save' button at the bottom. Use the 'Save' button on the top toolbar to continue saving your work throughout.

Step 5: Add basemaps
In the top toolbar left-click 'Open Data Source Manager'. From the left-hand tab select 'Raster'. Click on the '...' button to navigate to the GeoTIFF location. You can select multiple GeoTIFFs using the 'Shift' key on your keyboard and add them in all at once.

OR

Find 'XYZ tiles' in the Browser panel and use the arrow on the left to open the dropdown. Click and drag in OpenStreetMap or another WMS basemap from the Browser panel to the Layers panel.

Step 6: Zoom to map
Once the GeoTIFFs have loaded, right-click on one of the raster layers in the left-hand layers panel and click 'Zoom to layer'. Use the Pan and Zoom tools to find the location you want to map.

Step 7: Save Project
In the top toolbar choose 'Save Project As' and title QGIS Project with the site name. Use the 'Save' button on the top toolbar to continue saving your work throughout.

Step 8: Create New Layout
In the top panel click 'Create New Print Layout' button in the top toolbar.
Type a name for your layout in the pop-up box.

Step 9: Draw the map
In the Layout window, hover over the left-hand side toolbar until you come to 'Add new map'.

Click on 'Add new map' and click at the top left-hand corner of the blank page and drag a box across the page and double-click to finish. The map should appear on the screen.

Step 10: Type scale
In the 'Item properties' on the right-hand toolbar is a box containing the scale, if this does not appear, use the select tool to select the map you have added.

Type an appropriate scale for the map.

Step 11: Add scalebar
Go back to the left-hand side toolbar and find the 'Add new scalebar' button.

Click on 'Add new scalebar' then click on the map.

Ensure you have the Map selected when you add in the scalebar, or select 'Map 1' in the scale bar Item properties dropdown.

Step 12: Add copyright statement
Use the Add new label button and draw the position of the text box with the left mouse button on the canvas. Format the text under the Item properties tab.

Add text using the current year e.g. 2020: "Contains OS data © Crown copyright and database right (year)."

Step 13: Export image
Hover over the top toolbar until you come to 'Export as image'. Left-click and in the pop-up box choose where on your system and in which format. Alternatively, 'Export as pdf'. For export to be printed at A4 size, 300dpi is usually sufficient.

12.2   Survey maps

Step 1: Open New Project
Open QGIS Desktop. Click 'New Project' button in the top toolbar.

Step 2: Add site boundary
If you have been provided with a site boundary shapefile add this by navigating to the shapefile in the Browser panel and dragging it into the Layers panel.

Step 3: Zoom to site boundary
Zoom to the site boundary by right-clicking on the shapefile layer in the left-hand layers panel and left-clicking 'Zoom to layer'.

Step 4: Connect to online basemaps
Find 'XYZ tiles' in the Browser panel and use the arrow on the left to open the dropdown. Click and drag 'OpenStreetMap' into the Layers panel.

OR

Click on 'Open Data Source Manager' in the top toolbar.

Click the 'ArcGIS REST Server' tab from the left side of the window.

Click 'New'.

Name the connection.

In the URL box type: https://services.arcgisonline.com/ArcGIS/rest/services/World_Imagery/MapServer

Click 'OK'

Back in the Data Source Manager window click 'Connect'.

Select 'World Imagery' and click on 'Add'.

Wait for imagery layer to load.

Drag the boundary above the aerial layer.

OR

Once the connection is established the imagery is now available in the Browser panel. Click the arrow at the left side to show dropdown menu.

Double-click on 'World Imagery' to add to map.

Note: It is not recommended to add in all the other layers at once as the size of these tends to slow down or crash QGIS. Try one at a time.

Step 5: Create New Layout
In the top panel click 'Create New Print Layout' button in the top toolbar.

Type a name for your layout in the pop-up box.

Step 6: Draw the map
In the Layout window, hover over the left-hand side toolbar until you come to 'Add new map'.

Click on 'Add new map' and click at the top left-hand corner of the blank page and drag a box across the page and double-click to finish. The map should appear on the screen.

Step 7: Type scale
In the 'Item properties' on the right-hand toolbar is a box containing the scale, if this does not appear, use the select tool to select the map you have added.

Type an appropriate scale for the map.

Step 8: Add scalebar
Go back to the left-hand side toolbar and find the 'Add new scalebar' button. Click on 'Add new scalebar' then click on the map. Ensure you have the Map selected when you add in the scalebar, or select 'Map 1' in the scale bar Item properties dropdown.

Step 9: Add copyright statement
Use the Add new label button and draw the position of the text box with the left mouse button on the canvas. Format the text under the Item properties tab. Remember the copyright statement depends on the source of the basemap used:

Contains OS data © Crown copyright and database right (year)

(c) OpenStreetMap contributors

Source: Esri, Maxar, GeoEye, Earthstar Geographics, CNES/Airbus DS, USDA, USGS, AeroGRID, IGN, and the GIS User Community

Step 10: Add a north arrow
Left-click on the Add north arrow button and draw a box for the position you want for the north arrow. You should now see a north arrow where you have drawn the box. You can choose different arrows from the Item properties tab in the right toolbar.

Step 11: Add grid-lines
Click on the map and in the right-hand panel under 'Item properties', scroll down to 'Grid' and use the green plus button to add a new grid then click the 'Modify grid'...' button. In the next tab, change both the X and Y interval from 0 to 100. This gives you grid-lines at 100m separation (or use 1,000m for kilometre separation etc.). Scroll down and tick the 'Draw coordinates' box. Change the left and right formatting 'o 'Vertical ascending'. Adjust the size of your map on the page to ensure the grid coordinates will fit on the page.

Step 12: Export image
Click 'Export as image' in the top toolbar and choose where on your system and in which format.

12.3   Designated sites map

Step 1: Download basemaps
Go to the OS Open Data downloads page and download data for the area you are working in by Ordnance Survey grid square. To download maps suitable for regional scale basemap for display at 1:15,000–1:30,000 scale these are listed on the website as 'OS VectorMap District, Data type: Raster'. To find your area, use the map on the webpage. Tick the box to download and click through to the next page where you need to enter information. Await email and download data by following the link.

OR

Step 1: Connect to online basemaps
Find 'XYZ tiles' in the Browser panel and use the arrow on the left to open the dropdown. Click and drag 'OpenStreetMap' into the Layers panel.

Step 2: Download Designated Sites boundaries and unzip files
To do so go to the Nature Scot/Natural England/Data Map Wales/Open Data NI website and download the SSSI, SAC and SPA shapefiles.

Unzip the folders you have downloaded.

OR

Step 2: Set up connection to Designated Sites boundaries Web Map/Feature Services
Click on 'Open Data Source Manager' in the top toolbar.

Scotland: Click the 'WMS/WMTS' tab from the left side of the window.

Click 'New'.

Name the connection.

Type the url: https://ogc.nature.scot/geoserver/protectedareas/ows

There does not appear to be a WFS currently available for Scotland.

Back in the Data Source Manager window click 'Connect'.

Double-click to add in layers.

England: Click the 'ArcGIS REST Server' tab from the left side of the window.

Click 'New'.

Name the connection.

Type the url: https://services.arcgis.com/JJzESW51TqeY9uat/arcgis/rest/services/

Back in the Data Source Manager window click 'Connect'.

Double-click to add in layers.

Wales: Click either the 'WMS/WMTS' tab or 'WFS/OGC API Features' tab from the left side of the window.

Click 'New'.

Name the connection.

Type the url:

WMS: https://datamap.gov.wales/geoserver/ows

WFS: http://lle.gov.wales/services/wfs/nrw

Back in the Data Source Manager window click 'Connect'. Then double-click to add in layers.

Northern Ireland: Click the 'WMS/WMTS' tab from the left side of the window.

Click 'New'.

Name the connection.

Type the url: https://gis.epa.ie/geoserver/wms?service=WMS&request=getCapabilities &version=1.3.0

Back in the Data Source Manager window click 'Connect'.

Double-click 'Connect'.

Northern Ireland: In the URL paste host address: https://gis.epa.ie/geoserver/wms? service=WMS&request=getCapabilities&version=1.3.0

to add in layers

Step 3: Open New Project
Open QGIS Desktop. Click 'New Project' button in the top toolbar.

Step 4: Set the Coordinate System
In the top toolbar click on Project>Properties... then select 'CRS' from the left-hand tab.

Choose OSGB36 / British National Grid EPSG 27700.

To set this as default for all projects: in the top toolbar left-click Settings>Options... then select 'CRS' from the left-hand tab.

For work in Northern Ireland and Ireland use: TM65/Irish Grid EPSG 29902.

Step 5: Save Project
In the top toolbar choose 'Save Project As', navigate to where you want to save the project and type a file name. Click 'Save' button at the bottom. Use the 'Save' button on the top toolbar to continue saving your work throughout.

Step 6: Add site boundary
If you have been provided with a site boundary shapefile add this by left-clicking on 'Open Data Source Manager' in the top toolbar and selecting 'Vector' from the left-hand tab. Click on the '...' button to navigate to the shapefile location. Add in the shapefile or shapefiles you want in the map. These have the '*.shp' file type. To make these easier to find you can change the files displayed from 'All files' to 'ESRI shapefiles' in the bottom right-hand corner of the Windows Explorer/Finder pop-up box.

Step 7: Zoom to site boundary
Zoom to the site boundary by right-clicking on the shapefile layer in the left-hand layers panel and left-clicking 'Zoom to layer'.

Step 8: Add basemaps
In the top toolbar left-click 'Open Data Source Manager'. From the left-hand tab select 'Raster'. Click on the '...' button to navigate to the GeoTIFF location. You can select multiple GeoTIFFs using the 'Shift' key on your keyboard and add them in all at once.

OR

Find 'XYZ tiles' in the Browser panel and use the arrow on the left to open the dropdown. Click and drag 'OpenStreetMap' into the Layers panel.

Step 9: Add designated sites
Use 'Open Data Source Manager' in the top toolbar and select 'Vector' from the left-hand tab. Click on the '...' button to navigate to the shapefile location. Add in the SSSI, SAC, SPA, RAMSAR, LNR, NNR, etc. shapefiles. You can select multiple shapefiles using the 'Shift' key on your keyboard and add them in all at once.

Reorder layers in the layer tab with basemap below vector layers.

Save project.

OR

Open Data Source Manager

Click on the tab for the WMS connection you set up, select from the dropdown menu and then click 'Connect'.

Double-click to add layers or select and click 'Add'.

Close window.

Step 10: Use identify tool to find names of the designated sites
Find the 'Identify features' tool by hovering over the top toolbar.

This is a blue button with a white 'i'. Use it by clicking on the button, then selecting the layer you want to identify in the layers panel.

Then click on the designated site shapes one by one to identify them.

Step 11: Use the measure tool to measure the distance to designated sites
Find the 'measure' tool by hovering over the top toolbar. This is a picture of a ruler.

Use it by clicking on the button, then clicking on the site boundary and then the closest edge of the designated site.

Step 12: Symbology
To change the symbology of a vector layer, double-click a layer in the layers panel (or right-click and select 'Properties'). This opens the layer properties.

Choose the symbology tab from the left-hand side of the pop-up box.

Click on 'Simple fill' in the top box.

Choose 'Simple fill'.

Change the colour by choosing from the palette of standard colours.

Change fill style, stroke colour. Click OK to close the window.

Save project – This will save all changes.

Repeat for each designated site type, using a different colour for each. Often designations overlap each other for instance SSSIs underpin SACs. To show both at once you can show these both at once by changing the 'Solid fill' to one of the other options for lines or hatching. Alternatively, you can adjust the global opacity of the layer.

OR

Step 12: Transparency
To change the transparency of a raster layer, double-click a layer in the layers panel (or right-click and select 'Properties'). This opens the layer properties.

Choose the transparency tab from the left-hand side of the pop-up box.

Alter the 'Global opacity' using the slider or type a percentage into the box.

Click OK to close the window.

Step 13: Frame the map
Use the map navigation tools (Pan, zoom, zoom to layer etc.) so you can see all the vector layers in your map view.

Step 14: Create New Layout
In the top panel click 'Create New Print Layout' button in the top toolbar.

Type a name for your layout in the pop-up box.

Step 15: Draw the map
In the Layout window, hover over the left-hand side toolbar until you come to 'Add new map'.

Click on 'Add new map' and click at the top left-hand corner of the blank page and drag a box across the page and double-click to finish. The map should appear on the screen.

Step 16: Type scale
In the 'Item properties' on the right-hand toolbar is a box containing the scale, if this does not appear, use the select tool to select the map you have added.

Type an appropriate scale for the map.

Step 17: Add scalebar
Go back to the left-hand side toolbar and find the 'Add new scalebar' button.

Click on 'Add new scalebar' then click on the map.

Ensure you have the Map selected when you add in the scalebar, or select 'Map 1' in the scale bar Item properties dropdown.

Step 18: Add copyright statement
Use the Add new label button and draw the position of the text box with the left mouse button on the canvas. Format the text under the Item properties tab.

Add text:

"Contains OS data © Crown copyright and database right (year)"

AND

"© SNH, Contains Ordnance Survey data © Crown copyright and database right (2022)"
OR

"© Natural England copyright. Contains Ordnance Survey data © Crown copyright and database right [year]."
OR

"Contains Natural Resources Wales information © Natural Resources Wales and Database Right. All rights Reserved. Contains Ordnance Survey Data. Ordnance Survey Licence number 100019741. Crown Copyright and Database Right."

Step 19: Add a north arrow
Left-click on the Add north arrow button and draw a box for the position you want for the north arrow. You should now see a north arrow where you have drawn the box. You can choose different arrows from the Item properties tab in the right toolbar.

Step 20: Add a legend
Click the 'add new legend' button and draw a box for the legend. All legend entries will be added. Ensure legend is selected using the 'select item' button. Then modify as required under the 'item content' tab. First deselect auto update. Legend entries can then be added, removed, the order can be changed and the text can be edited using plus, minus, up and down buttons below.

Step 21: Export image
Hover over the top toolbar until you come to 'Export as image'. Left-click and in the pop-up box and choose where on your system and in which format. Alternatively, 'Export as pdf'. For export to be printed at A4 size 300dpi is usually sufficient.

12.4   Desk-study maps

Importing spreadsheet data

Step 1: Save spreadsheet data
Enter your data into Excel and save the sheet as a Common Separated Value (.csv) file.

In order to import data to plot in QGIS you need to add columns into your data for grid references. These can be as one column for British National Grid coordinates (e.g. NN705649) or two columns for Eastings and Northings (e.g. 270500, 764900).

OSGB grid references can be read off OS Maps in the field or recorded using a GPS.

OSGB grid references and Eastings and Northings are in metres and these are recommended for use in the UK.

To convert from Latitude and Longitude coordinates or Postcode use on option is to use Grid reference finder: gridreferencefinder.com. Please check license conditions allow your intended use.

Step 2: Open New Project
Open QGIS Desktop. Click 'New Project' button in the top toolbar.

Step 3: Connect to online basemaps
Click and drag in OpenStreetMap or another WMS basemap from the Browser panel to the Layers panel.

Step 4: Set the Coordinate System
In the top toolbar click on Project>Properties... then select 'CRS' from the left-hand tab.

Choose OSGB36 / British National Grid EPSG 27700.

To set this as default for all projects: in the top toolbar left-click Settings>Options... then select 'CRS' from the left-hand tab.

For work in Northern Ireland and Ireland use: TM65/Irish Grid EPSG 29902.

Step 5: Save Project
In the top toolbar choose 'Save Project As', navigate to where you want to save the project and type a file name. Click 'Save' button at the bottom. Use the 'Save' button on the top toolbar to continue saving your work throughout.

Step 6: Importing points using TomBio Plugin
If installed, when you right-click on the toolbars you should have the option to select the 'TomBio Toolbar' and a tick will appear beside it in the list. If it is not installed, install it from the Plugins tab.

On the TomBio Toolbar click the 'Biological records tool' which is a button with blue squares. Under the 'Data specification' tab, browse to the .csv file you want to import from the 'Create source from CSV file' dropdown menu.

To import the coordinates in British National Grid format, select the column containing the grid reference 'OSGR' from the OS Grid Ref Column pull down menu. Alternatively, you can import the coordinates as 'Eastings' and 'Northings' using the 'X' and 'Y' dropdown menus respectively.

Then click 'Create map layer' at the bottom of the TomBio plugin pop-up box, this is the same icon again with blue squares on. Depending on the size of your screen you may need to resize the TomBio plugin pop-up box.

Step 7: Save as shapefile
Once the new layer has been added to the layers panel, convert it to a shapefile. Right-click on the layer and select 'Save features as' Select 'ESRI Shapefile'. as the format and ensure that the CRS is British National grid. Tick 'Add' saved file to map'. Browse to a suitable folder and save the new file. Remove the original file from the map legend; right-click on the layer and select 'Remove'.

Step 8: Create a spatial index for the shapefile
In the top toolbar go to Vector>Data Management Tools>Create Spatial index.

Step 9: Style point data
To change the symbology of a layer, double-click a layer in the layers panel (or right-click and select 'Properties'). This opens the layer properties.

Choose the symbology tab from the left-hand side of the pop-up box.

Click on 'Simple marker' in the top box.

Choose 'Simple marker'.

Change the colour by choosing from the palette of standard colours.

Change fill style, stroke colour. Click OK to close the window.

Save project – This will save all changes.

Step 10: Frame the map
Use the map navigation tools (Pan, zoom, zoom to layer etc.) so you can see all the vector layers in your map view.

Step 11: Create New Layout
In the top panel click 'Create' New Print Layout' button in the top toolbar.

Type a name for your layout in the pop-up box.

Step 12: Draw the map
In the Layout window, hover over the left-hand side toolbar until you come to 'Add new map'.

Click on 'Add new map' and click at the top left-hand corner of the blank page and drag a box across the page and double-click to finish. The map should appear on the screen.

Step 13: Type scale
In the 'Item properties' on the right-hand toolbar is a box containing the scale, if this does not appear, use the select tool to select the map you have added.

Type an appropriate scale for the map.

Step 14: Add scalebar
Go back to the left-hand side toolbar and find the 'Add new scalebar' button.

Click on 'Add new scalebar' then click on the map.

Ensure you have the Map selected when you add in the scalebar, or select 'Map 1' in the scale bar Item properties dropdown.

Step 15: Add copyright statement
Use the Add new label button and draw the position of the text box with the left mouse button on the canvas. Format the text under the Item properties tab.

Add text copyright text for data and basemap.

Step 16: Add a north arrow
Left-click on the 'Add north arrow' button and draw a box for the position you want for the north arrow. You should now see a north arrow where you have drawn the box. You can choose different arrows from the 'Item properties' tab in the right toolbar.

Step 17: Add a legend
Click the 'add new legend' button and draw a box for the legend. All legend entries will be added. Ensure legend is selected using the 'select item' button. Then modify as required under the 'item content' tab. First deselect auto update. Legend entries can then be added, removed, the order can be changed and the text can be edited using plus, minus, up and down buttons below.

Step 18: Export image
Hover over the top toolbar until you come to 'Export as image' left-click and in the pop-up box and choose where on your system and in which format. Alternatively, 'Export as pdf'. For export to be printed at A4 size 300dpi is usually sufficient.

Importing data from NBN

Step 1: Download data from the NBN
Go to the NBN Atlas website and log in or create a log in. Search for dataset(s) you wish to download and look up the licence. Ask for permission to use the data (not required for Open Government Licence data). On receipt of permission download the data. This will be saved as a .csv file on your computer that you can then add in using the TomBio Plugin.

Step 2: Open New Project
Open QGIS Desktop. Click 'New Project' button in the top toolbar.

Step 3: Connect to online basemaps
Click and drag in OpenStreetMap or another WMS basemap from the Browser panel to the Layers panel.

Step 4: Set the Coordinate System
In the top toolbar click on Project>Properties... then select 'CRS' from the left-hand tab.

Choose OSGB36 / British National Grid EPSG 27700.

To set this as default for all projects: in the top toolbar left-click Settings>Options... then select 'CRS' from the left-hand tab.

For work in Northern Ireland and Ireland use: TM65/Irish Grid EPSG 29902.

Step 5: Save Project
In the top toolbar choose 'Save Project As', navigate to where you want to save the project and type a file name. Click 'Save' button at the bottom. Use the 'Save' button on the top toolbar to continue saving your work throughout.

Step 6: Importing atlas squares using TomBio Plugin
If installed, when you right-click on the toolbars you should have the option to select the 'TomBio Toolbar' and a tick will appear beside it in the list. If it is not installed, install it from the Plugins tab.

On the TomBio Toolbar click the 'Biological records tool' which is a button with blue squares. Under the 'Data specification' tab, browse to the .csv file you want to import from the 'Create' source from CSV file' dropdown menu.

To import the coordinates in British National Grid format, select the column containing the grid reference 'OSGR' from the OS Grid Ref Column pull down menu. Alternatively you can import the coordinates as 'Eastings' and 'Northings' using the 'X' and 'Y' dropdown menus respectively.

You can choose 'Scientific name' from the Taxon column dropdown menu to import one or more species. You can choose 'Abundance' to import an abundance field, this is useful for creating categorized symbols based on abundance rather than only displaying presence/absence.

Select '10km atlas (hectads from the 'Records as points' dropdown menu.

Then click 'Create' map layer' at the bottom of the TomBio plugin pop-up box, this is the same icon again with blue squares on. Depending on the size of your screen you may need to resize the TomBio plugin pop-up box.

Step 7: Save as shapefile

Once the new layer has been added to the layers panel, convert it to a shapefile. Right-click on the layer and select 'Export' then 'Save features as...' Select 'ESRI Shapefile' as the format and ensure that the CRS is British National grid. Tick 'Add' saved file to map'. Browse to a suitable folder and save the new file. Remove the original file from the map legend; right-click on the layer and select 'Remove'.

Step 8: Create a spatial index for the shapefile

In the top toolbar go to Vector>Data Management Tools>Create Spatial index.

Step 9: Symbology

To change the symbology of a layer, double-click a layer in the layers panel (or right-click and select 'Properties'). This opens the layer properties.

Choose the symbology tab from the left-hand side of the pop-up box.

Click on 'Simple fill' in the top box.

Choose 'Simple fill'.

Change the colour by choosing from the palette of standard colours.

Change fill style, stroke colour. Click OK to close the window.

OR

Step 9: Categorized symbology

Double-click a layer in the left layers panel or right-click and select 'Properties'.

Choose the 'Style' tab.

At the top where 'Single symbol' is automatically shown, click on the dropdown menu arrow to the right and select 'Categorized'.

Under 'Column' choose the field from the shapefile that you wish to categorize by.

Click 'Classify'. Arbitrary styles will be applied to each symbol. Double-click on each symbol and choose the style and colour you would like for each.

Save project – This will save all changes.

Step 10: Frame the map

Use the map navigation tools (Pan, zoom, zoom to layer etc.) so you can see all the vector layers in your map view.

Step 11: Create New Layout

In the top panel click 'Create New Print Layout' button in the top toolbar.

Type a name for your layout in the pop-up box.

Step 12: Draw the map

In the Layout window, hover over the left-hand side toolbar until you come to 'Add new map'.

Click on 'Add new map' and click at the top left-hand corner of the blank page and drag a box across the page and double-click to finish. The map should appear on the screen.

Step 13: Type scale
In the 'Item properties' on the right-hand toolbar is a box containing the scale, if this does not appear, use the select tool to select the map you have added.

Type an appropriate scale for the map.

Step 14: Add scalebar
Go back to the left-hand side toolbar and find the 'Add new scalebar' button.

Click on 'Add new scalebar' then click on the map.

Ensure you have the Map selected when you add in the scalebar, or select 'Map 1' in the scale bar Item properties dropdown.

Step 15: Add copyright statement
Use the 'Add new label' button and draw the position of the text box with the left mouse button on the canvas. Format the text under the Item properties tab.

Add text copyright text for data and basemap.

Step 16: Add a north arrow
Left-click on the Add north arrow button and draw a box for the position you want for the north arrow. You should now see a north arrow where you have drawn the box. You can choose different arrows from the Item properties tab in the right toolbar.

Step 17: Add a legend
Click the 'add new legend' button and draw a box for the legend. All legend entries will be added. Ensure legend is selected using the 'select item' button. Then modify as required under the 'item content' tab. First deselect auto update. Legend entries can then be added, removed, the order can be changed and the text can be edited using plus, minus, up and down buttons below.

Step 18: Export image
Hover over the top toolbar until you come to 'Export as image' left-click and in the pop-up box and choose where on your system and in which format. Alternatively, 'Export as' 'pdf'. For export to be printed at A4 size 300dpi is usually sufficient.

12.5   Protected species survey maps

Creating point data: importing from a GPS

Step 1: Open New Project
Open QGIS Desktop. Click 'New Project' button in the top toolbar.

Step 2: Download basemaps/Connect to online basemaps
Go to the OS Open Data downloads page and download data for your local area or the area you are working in by Ordnance Survey grid square. To the most detailed basemap to show at 1:20,000–1:30,000 scale is listed on the website as 'OS Open Map – Local, Data type: Raster'. To find your area, use the map on the webpage. Tick the box to download and click through to the next page where you need to enter information. Await email and download data by following the link.

Unzip the folders you have downloaded.

In the top toolbar left-click 'Open Data Source Manager'. From the left-hand tab select 'Raster'. Click on the '…' button to navigate to the GeoTIFF location. You can select multiple GeoTIFFs using the 'Shift' key on your keyboard and add them in all at once.

OR

Click and drag in OpenStreetMap or another WMS basemap from the Browser panel to the Layers panel.

Step 3: Set the Coordinate System
In the top toolbar click on Project>Properties... then select 'CRS' from the left-hand tab.

Choose OSGB36 / British National Grid EPSG 27700.

To set this as default for all projects: in the top toolbar left-click Settings>Options... then select 'CRS' from the left-hand tab.

For work in Northern Ireland and Ireland use: TM65/Irish Grid EPSG 29902.

Step 4: Save Project
In the top toolbar choose 'Save Project As', navigate to where you want to save the project and type a file name. Click 'Save' button at the bottom. Use the 'Save' button on the top toolbar to continue saving your work throughout.

Step 5: Check Batch GPS Importer Plugin
In the top toolbar go to Plugins>Manage and Install Plugins... In the 'All' tab use search bar to find 'Batch GPS Importer'.

Install and/or tick 'Batch GPS Importer' to enable plugin.

Step 6: Importing GPX files
In the top toolbar Vector menu> Batch GPS Importer>Batch GPS Importer.

Browse... to select the folder containing the GPX file.

Tick only those present from Waypoints, Track and/or Route.

Type in export layer name

Select Geometry type

Select projection/CRS

Step 7: Save as shapefile
Once the new layer has been added to the layers panel, convert it to a shapefile. Right-click on the layer and select 'Save features as…' and then select 'ESRI Shapefile'. as the format and ensure that the CRS is British National grid. Tick 'Add' saved file to map'. Browse to a suitable folder and save the new file. Remove the original file from the map legend; right-click on the layer and select 'Remove'.

Step 8: Create a spatial index for the shapefile
In the top toolbar go to Vector>Data Management Tools>Create Spatial index.

Step 9: Categorized symbology

Double-click a layer in the left layers panel or right-click and select 'Properties'

Choose the 'Style' tab.

At the top where 'Single symbol' is automatically shown, click on the dropdown menu arrow to the right and select 'Categorized'.

In 'Value' dropdown choose the field from the shapefile that you wish to categorize by.

Click 'Classify'. Arbitrary styles will be applied to each symbol. Double-click on each symbol and choose the style and colour you would like for each.

Step 10: Label

Select the label tab and select the column from the shapefile you wish to label by.

Step 11: Save project

Use the 'Save' button on the top toolbar to save changes.

Step 12: Frame the map

Right-click on the shapefile layer in the left layers panel and select 'Zoom to layer'.

OR

Use the map navigation tools (Pan, zoom, zoom to layer etc.) so you can see all the points layers in your map view.

Step 13: Create New Layout

In the top panel click 'Create' New Print Layout' button in the top toolbar.

Type a name for your layout in the pop-up box.

Step 14: Draw the map

In the Layout window, hover over the left-hand side toolbar until you come to 'Add new map'.

Click on 'Add new map' and click at the top left-hand corner of the blank page and drag a box across the page and double-click to finish. The map should appear on the screen.

Step 15: Type scale

In the 'Item properties' on the right-hand toolbar is a box containing the scale, if this does not appear, use the select tool to select the map you have added.

Type an appropriate scale for the map.

Step 16: Add scalebar

Go back to the left-hand side toolbar and find the 'Add new scalebar' button.

Click on 'Add new scalebar' then click on the map.

Ensure you have the Map selected when you add in the scalebar, or select 'Map 1' in the scale bar Item properties dropdown.

Step 17: Add copyright statement

Use the Add new label button and draw the position of the text box with the left mouse button on the canvas. Format the text under the Item properties tab.

Add text copyright text for data and basemap.

Step 18: Add a north arrow
Left-click on the 'Add north arrow' button and draw a box for the position you want for the north arrow. You should now see a north arrow where you have drawn the box. You can choose different arrows from the Item properties tab in the right toolbar.

Step 19: Add a legend
Click the 'Add new legend' button and draw a box for the legend. All legend entries will be added. Ensure legend is selected using the 'Select item' button. Then modify as required under the 'Item content' tab. First deselect auto update. Legend entries can then be added, removed, the order can be changed and the text can be edited using plus, minus, up and down buttons below.

Step 20: Export image
Hover over the top toolbar until you come to 'Export as image' left-click and in the pop-up box choose, choosing where on your system and in which format. Alternatively, 'Export as pdf'. For export to be printed at A4 size 300dpi is usually sufficient.

12.6   Georeferencing maps

Georeferencing using OS grid-lines

Step 1: Scan in map. Images to be georeferenced need to be saved as *.jpeg or *.jpg format. Try to flatten your field map as best as possible and scan it in as straight as possible within the scanner. If your scanner does not allow you to save directly to jpeg or jpg, use an image manipulation program or take a screenshot to save a copy of the image as *.jpeg or *.jpg once it is scanned in. Photographs of images taken on a mobile device can be used for georeferencing; however, these are often warped as it is very difficult to take a photograph straight into an image. If you only have a photograph of a map available, you will need to adjust the Transformation Settings to see what gives the best fit.

Step 2: Open New Project
Open QGIS Desktop. Click 'New Project' button in the top toolbar.

Step 3: Download basemaps/Connect to online basemaps
For greatest accuracy in georeferencing it is better to georeference to a downloaded raster image than an online basemap, but you may prefer to georeference the same image to itself in the case of a satellite image. However, be aware that distances measured will be less accurate on a georeferenced image and then more so using an image georeferenced to an online basemap because their native coordinate systems differ.

Go to the OS Open Data downloads page and download data for your local area or the area you are working in by Ordnance Survey grid square. To the most detailed basemap to show at 1:20,000–1:30,000 scale is listed on the website as 'OS Open Map – Local, Data type: Raster'. To find your area, use the map on the webpage. Tick the box to download and click through to the next page where you need to enter information. Await email and download data by following the link.

Unzip the folders you have downloaded.

In the top toolbar left-click 'Open Data Source Manager'. From the left-hand tab select 'Raster'. Click on the '...' button to navigate to the GeoTIFF location. You can select multiple GeoTIFFs using the 'Shift' key on your keyboard and add them in all at once.

OR

Click and drag in OpenStreetMap or another WMS basemap from the Browser panel to the Layers panel.

Step 3: Set the Coordinate System
In the top toolbar click on Project>Properties... then select 'CRS' from the left-hand tab.

Choose OSGB36 / British National Grid EPSG 27700.

To set this as default for all projects: in the top toolbar left-click Settings>Options... then select 'CRS' from the left-hand tab.

For work in Northern Ireland and Ireland use: TM65/Irish Grid EPSG 29902.

Step 4: Save Project
In the top toolbar choose 'Save Project As', navigate to where you want to save the project and type a file name. Click 'Save' button at the bottom. Use the 'Save' button on the top toolbar to continue saving your work throughout.

Step 5: Scan and save map
Scan in your map and save image as JPG.

Step 6: Load raster into Georeferencer
Open "Georeferencer" from the Layers or Raster top toolbar menu.

Click on the raster button and open your scanned in .jpg file in the Georeferencer window.

Select 'OSGB36 / British National Grid EPSG 27700' for the Coordinate Reference System (CRS).

Step 7: Add Control points by typing in Coordinates
Click on the 'Add point' button.

Click on the top left intersection of the grid-lines to create a control point. In the box up box type in the 'X/East' coordinate by reading off the coordinate from the top of the image.

In the box up box type in the 'Y/North' coordinate by reading off the coordinate from the left-hand side.

Repeat this at the right top, left bottom and right bottom corners of the map adding control points. The points appear in the table and you can edit, delete or move the points.

Step 8: Set Settings
Click on the 'Transformation settings' button.

In the pop-up box change the Transformation type to 'Polynomial 1' and specify the file name and location of the output raster. Set the target CRS to the same projection as the reference layer – British National Grid. Leave re-sampling on the default nearest neighbour.

Name the file with a different name. Tick the 'Load in QGIS' box when done.

Step 9: Start Georeferencing
In the Georeferencer window, click on the 'Start Georeferencing' button insert image. This complete the process, saves the georeferenced image as a GeoTIFF and adds the image into the Project.

Step 10: Check accuracy of Georeferencing
To check the accuracy of the georeferencing right-click on the raster layer>Properties>Transparency and reduce the 'Global Opacity' to 50%. You now see how well the georeferenced image matches the basemap.

If necessary, repeat the above in order to get a better fit.

Georeferencing without OS grid-lines

Step 1: Open New Project
Open QGIS Desktop. Click 'New Project' button in the top toolbar.

Step 2: Download basemaps/Connect to online basemaps
Go to the OS Open Data downloads page and download data for your local area or the area you are working in by Ordnance Survey grid square. To the most detailed basemap to show at 1:20,000 – 1:30,000 scale is listed on the website as 'OS Open Map – Local, Data type: Raster'. To find your area, use the map on the webpage. Tick the box to download and click through to the next page where you need to enter information. Await email and download data by following the link.

Unzip the folders you have downloaded.

In the top toolbar left-click 'Open Data Source Manager'. From the left-hand tab select 'Raster'. Click on the '...' button to navigate to the GeoTIFF location. You can select multiple GeoTIFFs using the 'Shift' key on your keyboard and add them in all at once.

OR

Click and drag in OpenStreetMap or another WMS basemap from the Browser panel to the Layers panel.

Step 3: Set the Coordinate System
In the top toolbar click on Project>Properties... then select 'CRS' from the left-hand tab.
Choose OSGB36 / British National Grid EPSG 27700.

To set this as default for all projects: in the top toolbar left-click Settings>Options... then select 'CRS' from the left-hand tab.

For work in Northern Ireland and Ireland use: TM65/Irish Grid EPSG 29902.

Step 4: Save Project
In the top toolbar choose 'Save Project As', navigate to where you want to save the project and type a file name. Click 'Save' button at the bottom. Use the 'Save' button on the top toolbar to continue saving your work throughout.

Step 5: Scan and save map
Scan in your map and save image as JPG.

Step 6: Load raster into Georeferencer
Open "Georeferencer" from the Layers or Raster top toolbar menu.

Click on the raster button and open your scanned in *.jpg file in the Georeferencer window.

Select 'OSGB36 / British National Grid EPSG 27700' for the Coordinate Reference System (CRS).

Step 7: Add Control points by clicking on a point on the map
Control points are how we tell the image to 'attach' to the basemap. Aim to use points at all four corners of the map. Look for intersections of roads or corners of buildings that you can find on the image you are georeferencing and in the basemap you are georeferencing to. Click on Add Point in the toolbar. Select a point on the image then click 'from map canvas' which will take you through to the basemap. Select the same point on the main map canvas.

Repeat this until you have 4 control points approximately in each corner of the map image.

The points appear in the table and you can edit, delete or move the points.

Step 8: Set Settings
Once you have enough points, go to Transformation settings. Choose 'Thin Plate Spline' for the transformation type. Set the target CRS to the same projection as the reference layer – British National Grid. Leave re-sampling on the default nearest neighbour.

Name the file with a different name. Tick the 'Load in QGIS' box when done.

Step 9: Start Georeferencing
In the Georeferencer window, click on the 'Start Georeferencing' button insert image. This complete the process, saves the Georeferenced image as a GeoTIFF and adds the image into the Project.

Step 10: Check accuracy of Georeferencing
To check the accuracy of the georeferencing right-click on the raster layer>Properties>Transparency and reduce the 'Global Opacity' to 50%. You now see how well the georeferenced image matches the basemap.

If necessary repeat the above in order to get a better fit.

This can take a lot of practice to get proficient with and the accuracy will always be limited especially when georeferencing a scanned jpg with a different basemap or at a different scale.

12.7   Habitat survey maps

Basic Habitat Map

To georeference a survey map, follow the steps in Chapter 10 Georeferencing Maps.

Step 1: Open New Project
Open QGIS Desktop. Click 'New Project' button in the top toolbar.

Step 2: Download basemaps/Connect to online basemaps
Go to the OS Open Data downloads page and download data for your local area or the area you are working in by Ordnance Survey grid square. To the most detailed basemap to show at 1:20,000-1:30,000 scale is listed on the website as 'OS Open Map – Local, Data type: Raster'. To find your area, use the map on the webpage. Tick the box to download and click through to the next page where you need to enter information. Await email and download data by following the link.

Unzip the folders you have downloaded.

In the top toolbar left-click 'Open Data Source Manager'. From the left-hand tab select 'Raster'. Click on the '...' button to navigate to the GeoTIFF location. You can select multiple GeoTIFFs using the 'Shift' key on your keyboard and add them in all at once.

OR

For OpenStreetMap:

'OSM Standard' is the default that you can add in from the Browser panel>XYZ Tiles>OpenStreetMap.

OR

For satellite:

If you have a connection to aerial imagery from Chapter 6.4 Connecting to online basemaps you can add in World Imagery from the Browser panel>ArcGIS REST Servers>World Imagery>1.2m Resolution.

OR

To set up a new connection:

Open a web browser and search for World Imagery. At time of writing the website is: https://www.arcgis.com/home/item.html?id=10df2279f9684e4a9f6a7f08febac2a9

Back in QGIS click on 'Open Data Source Manager' in the left toolbar.

Scroll down and Click the 'ArcGIS REST Server' tab from the left side of the window.

Click 'New'. Give a name for the connection, in this case 'World Imagery'.

In the URL box paste in the URL you copied from the website. At time of writing this is: https://services.arcgisonline.com/ArcGIS/rest/services/World_Imagery/MapServer Click 'OK'.

Back in the Data Source Manager window click 'Connect'.

Select 'World Imagery' and click on the Raster layer '1.2m Resolution Metadata' then 'Add'.

Step 3: Set the Coordinate System
In the top toolbar click on Project>Properties... then select 'CRS' from the left-hand tab.

Choose OSGB36 / British National Grid EPSG 27700.

To set this as default for all projects: in the top toolbar left-click Settings>Options... then select 'CRS' from the left-hand tab.

For work in Northern Ireland and Ireland use: TM65/Irish Grid EPSG 29902.

Step 4: Save Project
In the top toolbar choose 'Save Project As', navigate to where you want to save the project and type a file name. Click 'Save' button at the bottom. Use the 'Save' button on the top toolbar to continue saving your work throughout.

Step 5: Load in Georeferenced survey map image
In the top toolbar left-click 'Open Data Source Manager'. From the left-hand tab select 'Raster'. Click on the '...' button to navigate to the GeoTIFF location. You can select multiple GeoTIFFs using the 'Shift' key on your keyboard and add them in all at once.

Step 6: Create Point, Line and Polygon shapefiles

Select 'New Shapefile Layer' from the top toolbar layer menu.

In the pop-up box on the furthest right-hand side after 'File name', use the '…' button to navigate to where you want to save the shapefile and type a file name.

For 'Geometry type' use the dropdown arrow and select 'Point', 'Line' or 'Polygon'.

Check the CRS is OSGB36 / British National Grid EPSG 27700 or for work in Northern Ireland and Ireland use: TM65/Irish Grid EPSG 29902.

Under 'New field>Name' type "Habitat", leave the 'Type' as 'Text data' and change 'Length' from "80" to "200". Click the 'Add to fields list' button.

Then 'OK' at the bottom to create your shapefile.

Step 7: Digitizing

Note: When digitizing from a georeferenced image, be aware that how good a fit you got on the georeferencing will affect the accuracy of where the image is displaying on the screen. This means that features on the image you are digitizing may be shifted away from where they really are. Therefore, you should digitize from the basemap rather than the image if they is disagreement on the exact position of feature. For instance, if you are digitizing a field with a road on one side and the image is shifted a few centimetres to the left of the road on the basemap, you should assume the basemap is correct and digitize to the basemap road for the boundary. You also need to bear this in mind for the accuracy you quote to for any measurements taken from digitized features.

Select a shapefile layer by left-clicking on it in the left-hand layers panel and choose the 'toggle editing' button from the editing toolbar. This starts and stops editing.

The 'add features' button will put the cursor in digitizing mode. Left-click on the map area to create the first point of your new feature. Keep clicking for each additional point.

Once you have finished right-click. The attribute window will appear to allow you to enter information for that feature.

Save your layer edits regularly! You can either use the 'save edits' button or when you toggle off editing, you will be prompted to save changes.

Step 8: Categorized symbology

Double-click a layer in the left layers panel or right-click and select 'Properties'.

Choose the 'Style' tab.

At the top where 'Single symbol'/'Simple line'/'Single marker' is automatically shown, click on the dropdown menu arrow to the right and select 'Categorized'.

Under 'Column' choose the 'Habitat' field from the shapefile to categorize by.

Click 'Classify'. Arbitrary styles will be applied to each symbol. Double-click on each symbol and choose the style and colour you would like for each.

Step 9: Frame the map

Use the map navigation tools (Pan, zoom, zoom to layer etc.) so you can see all the vector layers in your map view.

Step 10: Create New Layout

In the top panel click 'Create New Print Layout' button in the top toolbar.

Type a name for your layout in the pop-up box.

Step 11: Draw the map
In the Layout window, hover over the left-hand side toolbar until you come to 'Add new map'.

Click on 'Add new map' and click at the top left-hand corner of the blank page and drag a box across the page and double-click to finish. The map should appear on the screen.

Step 12: Type scale
In the 'Item properties' on the right-hand toolbar is a box containing the scale, if this does not appear, use the select tool to select the map you have added.

Type an appropriate scale for the map.

Step 13: Add scalebar
Go back to the left-hand side toolbar and find the 'Add new scalebar' button.

Click on 'Add new scalebar' then click on the map.

Ensure you have the Map selected when you add in the scalebar, or select 'Map 1' in the scale bar Item properties dropdown.

Step 14: Add copyright statement
Use the Add new label button and draw the position of the text box with the left mouse button on the canvas. Format the text under the Item properties tab.
Add text copyright text for survey data and basemap.

Step 15: Add a north arrow
Left-click on the 'Add north arrow' button and draw a box for the position you want for the north arrow. You should now see a north arrow where you have drawn the box. You can choose different arrows from the Item properties tab in the right toolbar.

Step 16: Add a legend
Click the 'Add new legend' button and draw a box for the legend. All legend entries will be added. Ensure legend is selected using the 'Select item' button. Then modify as required under the 'Item content' tab. First deselect auto update. Legend entries can then be added, removed, the order can be changed and the text can be edited using plus, minus, up and down buttons below.

Step 17: Export image
Hover over the top toolbar until you come to 'Export as image' click and in the pop-up box choose, where on your system and in which format. Alternatively, 'Export as pdf'. For export to be printed at A4 size 300dpi is usually sufficient.

Intermediate Habitat Map

To Georeference a survey map, follow the steps in 13.6 Georeferencing maps above.

Step 1: Open New Project
Open QGIS Desktop. Click 'New Project' button in the top toolbar.

Step 2: Download basemaps/Connect to online basemaps

Go to the OS Open Data downloads page and download data for your local area or the area you are working in by Ordnance Survey grid square. To the most detailed basemap to show at 1:20,000–1:30,000 scale is listed on the website as 'OS Open Map – Local, Data type: Raster'. To find your area, use the map on the webpage. Tick the box to download and click through to the next page where you need to enter information. Await email and download data by following the link.

Unzip the folders you have downloaded.

In the top toolbar left-click 'Open Data Source Manager'. From the left-hand tab select 'Raster'. Click on the '...' button to navigate to the GeoTIFF location. You can select multiple GeoTIFFs using the 'Shift' key on your keyboard and add them in all at once.

OR

For OpenStreetMap:

'OSM Standard' is the default that you can add in from the Browser panel>XYZ Tiles>OpenStreetMap.

OR

For satellite:

If you have a connection to aerial imagery from Chapter 6.4 Connecting to online basemaps you can add in World Imagery from the Browser panel>ArcGIS REST Servers>World Imagery>1.2m Resolution.

OR

To set up a new connection:

Open a web browser and search for World Imagery. At time of writing, the website is: https://www.arcgis.com/home/item.html?id=10df2279f9684e4a9f6a7f08febac2a9

Back in QGIS click on 'Open Data Source Manager' in the left toolbar.

Scroll down and Click the 'ArcGIS REST Server' tab from the left side of the window.

Click 'New'. Give a name for the connection, in this case 'World Imagery'.

In the URL box paste in the URL you copied from the website, at time of writing this is: https://services.arcgisonline.com/ArcGIS/rest/services/World_Imagery/MapServer Click 'OK'.

Back in the Data Source Manager window click 'Connect'.

Select 'World Imagery' and click on the Raster layer '1.2m Resolution Metadata' then 'Add'.

Step 3: Set the Coordinate System

In the top toolbar click on Project>Properties... then select 'CRS' from the left-hand tab.

Choose OSGB36 / British National Grid EPSG 27700.

To set this as default for all projects: in the top toolbar left-click Settings>Options...then select 'CRS' from the left-hand tab.

For work in Northern Ireland and Ireland use: TM65/Irish Grid EPSG 29902.

Step 4: Save Project
In the top toolbar choose 'Save Project As', navigate to where you want to save the project and type a file name. Click 'Save' button at the bottom. Use the 'Save' button on the top toolbar to continue saving your work throughout.

Step 5: Load in Georeferenced survey map image
In the top toolbar left-click 'Open Data Source Manager'. From the left-hand tab select 'Raster'. Click on the '...' button to navigate to the GeoTIFF location. You can select multiple GeoTIFFs using the 'Shift' key on your keyboard and add them in all at once.

Step 6: Create Point, Line and Polygon shapefiles
Select 'New Shapefile Layer' from the top toolbar layer menu.

In the pop-up box on the far right-hand side after 'File name', use the '...' button to navigate to where you want to save the shapefile and type a file name.

For 'Geometry type' use the dropdown arrow and choose 'Point', 'Line' or 'Polygon'.

Check the CRS is OSGB36 / British National Grid EPSG 27700 or for work in Northern Ireland and Ireland use: TM65/Irish Grid EPSG 29902.

Under 'New field>Name' type "Habitat", leave the 'Type' as 'Text data' and change 'Length' from "80" to "200". Click the 'Add to fields list' button.

Then 'OK' at the bottom to create your shapefile.

Step 7: Digitizing
Select a shapefile layer by left-clicking on it in the left-hand layers panel and choose the 'Toggle editing' button from the editing toolbar. This starts and stops editing.

The 'Add features' button will put the cursor in digitizing mode. Left-click on the map area to create the first point of your new feature. Keep clicking for each additional point.

Once you have finished right-click. The attribute window will appear to allow you to enter information for that feature.

Save your layer edits regularly! You can either use the 'Save edits' button or when you toggle off editing, you will be prompted to save changes.

Step 8: Basic edits
Select layer in Layers panel. Use the Select feature button to select then Delete a feature.

Right-click in the top toolbar and turn on 'Snapping toolbar'.

Turn on snapping by clicking on the magnet button.

To snap to vertices:

Use the second dropdown menu under the 3 square button to select 'Vertex'.

Set tolerance. This is the distance QGIS uses to search for the closest vertex (10 pixels usually works well). Choose pixels as the unit as then snapping stays the same irrespective of zoom.

Use the Vertex tool to move vertices:
Select layer in Layers panel. Select feature. Toggle editing. Select Vertex Tool.

Click on the vertex you want to move then click where you want it to move to.

Save your layer edits regularly! You can either use the 'save edits' button or when you toggle off editing, you will be prompted to save changes.

Step 9: Add style files
In your polygon habitats layer: Open Attribute table and use Add Field to add a String field.

The string must be called "p1code" to automatically link to this as the value field for the symbology from the style file provided. Character length of 10 is more than enough for all the habitat codes. Check attribute table for your new field.

In layer properties load polygon style file. This can be downloaded from Field Studies Council, from this webpage at time of writing: https://www.fscbiodiversity.uk/qgis-mapping-styles-uk-habitats

Type in the Phase 1 habitat code to the new field cell, ensure this is correct.

The style file will then be available for all future projects.

Repeat for the line habitats and target notes style files.

UK Habitat Classification has recently been published as a replacement for the Phase 1 Habitat survey. It specifies what size of habitat should be mapped as polygon, line and point. For more information consult http://ecountability.co.uk/ukhabworkinggroup-ukhab/

Step 10: Calculating areas
Use the field calculator button in the attribute table of your polygon shapefile.

Select Create a new field, type an output field name "Area".

Select Whole number (integer) from the Output field type.

Under functions, select Geometry.

Double-click $area. Click OK.

A new column will be added to the attribute table with the area calculated in metres squared.

Step 11: Frame the map
Use the map navigation tools (Pan, zoom, zoom to layer etc.) so you can see all the vector layers in your map view.

Step 12: Create New Layout
In the top panel click 'Create New Print Layout' button in the top toolbar.

Type a name for your layout in the pop-up box.

Step 13: Draw the map
In the Layout window, hover over the left-hand side toolbar until you come to 'Add new map'.

Click on 'Add new map' and click at the top left-hand corner of the blank page and drag a box across the page and double-click to finish. The map should appear on the screen.

Step 14: Type scale
In the 'Item properties' on the right-hand toolbar is a box containing the scale, if this does not appear, use the select tool to select the map you have added.

Type an appropriate scale for the map.

Step 15: Add scalebar
Go back to the left-hand side toolbar and find the 'Add new scalebar' button.

Click on 'Add new scalebar' then click on the map.

Ensure you have the Map selected when you add in the scalebar, or select 'Map 1' in the scale bar Item properties dropdown.

Step 16: Add copyright statement
Use the 'Add new label' button and draw the position of the text box with the left mouse button on the canvas. Format the text under the 'Item properties' tab.

Add text copyright text for survey data and basemap.

Step 17: Add a north arrow
Left-click on the Add north arrow button and draw a box for the position you want for the north arrow. You should now see a north arrow where you have drawn the box. You can choose different arrows from the Item properties tab in the right toolbar.

Step 18: Add a legend
Click the 'Add new legend' button and draw a box for the legend. All legend entries will be added. Ensure legend is selected using the 'Select item' button. Then modify as required under the 'Item content' tab. First deselect auto update. Legend entries can then be added, removed, the order can be changed and the text can be edited using plus, minus, up and down buttons below.

Step 19: Export image
Hover over the top toolbar until you come to 'Export as image' left-click and in the pop-up box choose where on your system and in which format. Alternatively, 'Export as pdf'. For export to be printed at A4 size 300dpi is usually sufficient.

Advanced Habitat Map

To Georeference a survey map, follow the steps in 13.6 Georeferencing maps above.

Step 1: Open New Project
Open QGIS Desktop. Click 'New Project' button in the top toolbar.

Step 2: Download basemaps/Connect to online basemaps
Go to the OS Open Data downloads page and download data for your local area or the area you are working in by Ordnance Survey grid square. To the most detailed basemap to show at 1:20,000–1:30,000 scale is listed on the website as 'OS Open Map – Local, Data type: Raster'. To find your area, use the map on the webpage. Tick the box to download and click through to the next page where you need to enter information. Await email and download data by following the link.

Unzip the folders you have downloaded.

In the top toolbar left-click 'Open Data Source Manager'. From the left-hand tab select 'Raster'. Click on the '...' button to navigate to the GeoTIFF location. You can select multiple GeoTIFFs using the 'Shift' key on your keyboard and add them in all at once.

OR

For OpenStreetMap:

'OSM Standard' is the default that you can add in from the Browser panel>XYZ Tiles>OpenStreetMap.

OR

For satellite:

If you have a connection to aerial imagery from Chapter 6.4 Connecting to online basemaps you can add in World Imagery from the Browser panel>ArcGIS REST Servers>World Imagery>1.2m Resolution.

OR

To set up a new connection:

Open a web browser and search for World Imagery. At time of writing the website is: https://www.arcgis.com/home/item.html?id=10df2279f9684e4a9f6a7f08febac2a9

Back in QGIS click on 'Open Data Source Manager' in the left toolbar.

Scroll down and Click the 'ArcGIS REST Server' tab from the left side of the window.

Click 'New'. Give a name for the connection, in this case 'World Imagery'.

In the URL box paste in the URL you copied from the website, at time of writing this is:

https://services.arcgisonline.com/ArcGIS/rest/services/World_Imagery/MapServer

Click 'OK'.

Back in the Data Source Manager window click 'Connect'.

Select 'World Imagery' and click on the Raster layer '1.2m Resolution Metadata' then 'Add'.

Step 3: Set the Coordinate System
In the top toolbar click on Project>Properties... then select 'CRS' from the left-hand tab.

Choose OSGB36 / British National Grid EPSG 27700.

To set this as default for all projects: in the top toolbar left-click Settings>Options... then select 'CRS' from the left-hand tab.

For work in Northern Ireland and Ireland use: TM65/Irish Grid EPSG 29902

Step 4: Save Project
In the top toolbar choose 'Save Project As', navigate to where you want to save the project and type a file name. Click 'Save' button at the bottom. Use the 'Save' button on the top toolbar to continue saving your work throughout.

Step 5: Load in Georeferenced survey map image
In the top toolbar left-click 'Open Data Source Manager'. From the left-hand tab select 'Raster'. Click on the '...' button to navigate to the GeoTIFF location. You can select multiple GeoTIFFs using the 'Shift' key on your keyboard and add them in all at once.

Step 6: Create Point, Line and Polygon shapefiles
Select 'New Shapefile Layer' from the top toolbar layer menu.

In the pop-up box on the furthest right-hand side after 'File name', use the '...' button to navigate to where you want to save the shapefile and type a file name.

For 'Geometry type' use the dropdown arrow and choose 'Point', 'Line' or 'Polygon'.

Check the CRS is OSGB36 / British National Grid EPSG 27700 or for work in Northern Ireland and Ireland use: TM65/Irish Grid EPSG 29902.

Under 'New field>Name' type "Habitat", leave the 'Type' as 'Text data' and change 'Length' from "80" to "200". Click the 'Add to fields list' button.
Then 'OK' at the bottom to create your shapefile.

Step 7: Digitizing
Select a shapefile layer by left-clicking on it in the left-hand layers panel and choose the toggle editing button from the editing toolbar. This starts and stops editing.

The 'Add features' button will put the cursor in digitizing mode. Left-click on the map area to create the first point of your new feature. Keep clicking for each additional point.

Once you have finished right-click. The attribute window will appear to allow you to enter information for that feature.

Save your layer edits regularly! You can either use the 'save edits' button or when you toggle off editing, you will be prompted to save changes.

Step 8: Basic edits
Use the Select feature button to select then Delete a feature.

Right-click in the top toolbar and turn on 'Snapping toolbar'.

Turn on snapping by clicking on the 'Snapping' button.

To snap to vertices:

Use the second dropdown menu in the 'Snapping toolbar' to select 'Vertex'.

Set tolerance. Select px (pixels) as the unit and 10 as the number. This will ensure the snapping tolerance is the same irrespective of zoom.

Use the 'Vertex tool' to move vertices:

Select by clicking one vertex at a time or clicking and dragging around several at a time. They will turn blue.

Add a vertex by double-clicking the edge of the polygon. Delete a vertex by selecting and pressing the delete key. Move by selecting the vertex and then clicking on the place you want it to move to. Don't try dragging it around by holding your mouse button down!

Save your layer edits regularly! You can either use the 'Save edits' button or, when you toggle off editing, you will be prompted to save changes.

Step 9: Advanced edits
Right-click the tool menu and select advanced digitizing to load the advanced digitizing toolbar.

First highlight the features you want to edit using 'Select features'.

Use the 'Add ring' button to add a hole into another polygon.

If you need to delete the ring and start again, select the 'Delete ring' button. Click inside ring and delete.

To add a polygon within another without an existing hole use the 'Fill Ring' tool.

To cut existing polygons into two, use the 'Split feature' tool to draw a line across the polygon where you want to split. Start outside the polygon, cross it and right-click outside. Two polygons will now be created.

Use the 'Reshape tool' to add to polygons: Click inside polygon, cross the boundary, add any vertices and then finish by right-clicking inside the polygon. To remove part of the polygon: do the reverse. Start outside, cross the boundary, add any vertices inside the polygon and right-click outside the polygon.

Use the 'Reshape' tool and 'Split features' tool do finer edits to the shapefiles.

Step 10: Add style files
In your polygon habitats layer: click 'Open Attribute table' and use 'Add Field' to add a 'String' field called "p1code" of length 10. Check attribute table for your new field.

In layer properties, load polygon style file from QGIS-UK-habitat-styles folder.

Type in the Phase 1 habitat code to the 'p1code' field for the symbology to automatically use the Phase 1 habitats styles.

The style file will then be available for all future projects.

UK Habitat Classification has recently been published as a replacement for the Phase 1 Habitat survey. It specifies what size of habitat should be mapped as polygon, line and point. For more information consult http://ecountability.co.uk/ukhabworkinggroup-ukhab/

Step 11: Calculating areas
Use the field calculator button in the attribute table of your polygon shapefile.

Select Create a new field, type an output field name 'Area'.

Select Whole number (integer) from the Output field type.

Under functions, select Geometry.

Double-click $area. Click OK.

A new column will be added to the attribute table with the area calculated in metres squared.

Step 12: Frame the map
Use the map navigation tools (Pan, zoom, zoom to layer etc.) so you can see all the vector layers in your map view.

Step 13: Create New Layout
In the top panel click 'Create New Print Layout' button in the top toolbar.
Type a name for your layout in the pop-up box.

Step 14: Draw the map
In the Layout window, hover over the left-hand side toolbar until you come to 'Add new map'.

Click on 'Add new map' and click at the top left-hand corner of the blank page and drag a box across the page and double-click to finish. The map should appear on the screen.

Step 15: Type scale
In the 'Item properties' on the right-hand toolbar is a box containing the scale, if this does not appear, use the select tool to select the map you have added.

Type an appropriate scale for the map.

Step 16: Add scalebar
Go back to the left-hand side toolbar and find the 'Add new scalebar' button.

Click on 'Add new scalebar' then click on the map.

Ensure you have the Map selected when you add in the scalebar, or select 'Map 1' in the scale bar Item properties dropdown.

Step 17: Add copyright statement
Use the Add new label button and draw the position of the text box with the left mouse button on the canvas. Format the text under the 'Item properties' tab.

Add text copyright text for survey data and basemap.

Step 18: Add a north arrow
Left-click on the Add north arrow button and draw a box for the position you want for the north arrow. You should now see a north arrow where you have drawn the box. You can choose different arrows from the Item properties tab in the right toolbar.

Step 19: Add a legend
Click the 'Add new legend' button and draw a box for the legend. All legend entries will be added. Ensure legend is selected using the 'Select item' button. Then modify as required under the 'Item content' tab. First deselect auto update. Legend entries can then be added, removed, the order can be changed and the text can be edited using plus, minus, up and down buttons below.

Step 20: Export image
Hover over the top toolbar until you come to 'Export as image' click and in the pop-up choose where on your system and in which format. Alternatively, 'Export as pdf'. For export to be printed at A4 size 300dpi is usually sufficient.

Glossary

Coordinate	The letters and/or numbers that describe geographical location of a point
Coordinate Reference System (CRS)	The type of letters and/or numbers format that tells QGIS where to plot coordinates at a country or world scale.
Desk study	Collating of existing site data prior to field survey
Designated site	An area of habitat protected by law
Digitize	To draw shapes, lines or points in QGIS
Geographical Information Systems (GIS)	Mapping software used for cartography and spatial analysis
Global Positioning System (GPS) device	Handheld electronic device for navigating using satellites
Global Positioning System (GPS)	Network of navigation satellites
Georeference	To add spatial information to an image file
GeoTIFF	Collection of files that together containing a map image that displays in the correct place geographically
Phase 1 Habitat Survey	A method of mapping habitats with colours and symbols
Plugin	Additional tool that can be added into QGIS
Protected Species Survey	A method of cataloguing animals and plants that are protected by law
Quantum Geographical Information System (QGIS)	A free open source mapping software used for cartography and spatial analysis
Raster	A spatial imagery layer that is not editable
Shapefile	Collection of files that together containing a shape that displays in the correct place geographically
Snap	To 'magnetize' the mouse cursor to vertexes, intersections or lines
UK Habitats Classification	Updated Phase 1 Habitats Survey methodology of mapping habitats with colours and symbols with guidance on size of features to be mapped
Vector	A spatial shape layer that is editable and has information about the shapes stored within it
Vertex	An editable point within a line or polygon shape
Workflow	Step-by-step process to follow

Data copyright and licenses

The aim of this book is to provide an educational resource for those working in or aiming to work in the ecological sector. To do so I have provided resources that to the best of my knowledge are free to use and publish both for educational and commercial purposes. The licenses and sources of these resources are listed in here.

Chapter	Data	License	Source	Restrictions	Copyright Statement
1 - 15	QGIS software	General Public License (GNU) v3.0 https://www.gnu.org/ licenses/gpl-3.0.en. html#license-text	The QGIS project: Downloaded from: https://qgis.org/en/site/	There is no requirement to mention QGIS for maps produced with it. It is of course really welcome if would like to add a note saying that the map was produced with QGIS. "Made with QGIS" or "Map created using the Free and Open Source QGIS" are good examples of such a note.	QGIS Development Team (2023). QGIS Geographic Information System. Open Source Geospatial Foundation Project. http://qgis.osgeo.org
5	Collymoon_SSSI_ boundary.shp	Open Government Licence (OGL) v3.0 https://www. nationalarchives.gov.uk/doc/ open-government-licence/ version/3/	Derived from Nature Scot (Scottish Natural Heritage) Sites of Special Scientific Interest (SSSI) shapefile downloaded from: https:// opendata.nature.scot/	Acknowledge the source of the Information in your product or application by including or linking to any attribution statement specified by the Information Provider(s) and, where possible, provide a link to this licence.	© SNH, Contains Ordnance Survey data © Crown copyright and database right (2022)
5	NS59NE.TIF	Open Government Licence (OGL) v3.0 https://www. nationalarchives.gov.uk/doc/ open-government-licence/ version/3/	OS Open Map Local Ordnance Survey Downloaded from: https://osdatahub.os.uk/ downloads/open	Acknowledge the source of the Information in your product or application by including or linking to any attribution statement specified by the Information Provider(s) and, where possible, provide a link to this licence.	Contains OS data © Crown copyright and database right (2023)

Chapter	Data	License	Source	Restrictions	Copyright Statement
6–11	OpenStreetMap	Open Data Commons Open Database License (ODbL)	OpenStreetMap https://www.openstreetmap.org WMS assessed via QGIS	You are free to copy, distribute, transmit and adapt our data, as long as you credit OpenStreetMap and its contributors. If you alter or build upon our data, you may distribute the result only under the same licence.	© OpenStreetMap contributors
6, 10 & 11	World Imagery	Esri World Imagery Map Terms of Use	Available at: https://www.arcgis.com/home/item.html?id=8e90a00a6845a49262e0b756f57a10 WMS assessed via QGIS	The World Imagery map may be used in various ArcGIS applications to support data collection and editing. Esri and its imagery contributors grant Users the non-exclusive right to use the World Imagery map to trace features and validate edits in the creation of vector data. Users that create vector data from the World Imagery map may want to publicly share that vector data through a GIS data clearinghouse of its own or through another open data site. This public sharing could be achieved through ArcGIS Open Data or the OpenStreetMap (OSM) Initiative. For ArcGIS users that want to contribute such vector data to OSM, Esri provides applications and services directly accessible from ArcGIS platform. Users acknowledge that any vector data contributed to OSM is then governed by and released under the OpenStreetMap License (e.g. ODbL). Except for the additional limited rights granted above, any and all other uses of the World Imagery map remain subject to the terms and conditions set forth in the Esri Master Agreement or Terms of Use, as applicable. Esri and its imagery contributors retain all right, title, and interest in and to their respective imagery data contributed to the World Imagery map.	Source: Esri, Maxar, Geo-ye, Earthstar Geographics, CNES/Airbus DS, USDA, USGS, AeroGRID, IGN, and the GIS User Community

Chapter	Data	License	Source	Restrictions	Copyright Statement
7	Edinburgh_Castle_boundary.shp	Open Data Commons Open Database License (ODbL)	Digitized from: OpenStreetMap https://www.openstreetmap.org WMS assessed via QGIS	You are free to copy, distribute, transmit and adapt our data, as long as you credit OpenStreetMap and its contributors. If you alter or build upon our data, you may distribute the result only under the same licence.	© OpenStreetMap contributors
7	Scotland Sites of Special Scientific Interest (SSSI) SSSI_SCOTLAND.shp Special Areas of Conservation (SACs) and Special Protection Areas (SPAs)	Open Government Licence (OGL) v3.0 https://www.nationalarchives.gov.uk/doc/open-government-licence/version/3/	Nature Scot (Scottish Natural Heritage) shapefile/ API downloaded from: https://opendata.nature.scot/ WMS assessed via QGIS	Acknowledge the source of the Information in your product or application by including or linking to any attribution statement specified by the Information Provider(s) and, where possible, provide a link to this licence.	© SNH, Contains Ordnance Survey data © Crown copyright and database right (2022)
7	England Sites of Special Scientific Interest (SSSI), Special Areas of Conservation (SACs) and Special Protection Areas (SPAs)	Open Government Licence (OGL) v3.0 https://www.nationalarchives.gov.uk/doc/open-government-licence/version/3	Natural England shapefile/ API downloaded from: https://naturalengland-defra.opendata.arcgis.com/ WMS assessed via QGIS	Acknowledge the source of the Information in your product or application by including or linking to any attribution statement specified by the Information Provider(s) and, where possible, provide a link to this licence.	© Natural England copyright. Contains Ordnance Survey data © Crown copyright and database right [2023].
7	Wales Sites of Special Scientific Interest (SSSI), Special Areas of Conservation (SACs) and Special Protection Areas (SPAs)	Open Government Licence (OGL) v3.0 https://www.nationalarchives.gov.uk/doc/open-government-licence/version/3	Data Map Wales shapefile/ API downloaded from: https://datamap.gov.wales/ WMS assessed via QGIS	Acknowledge the source of the Information in your product or application by including or linking to any attribution statement specified by the Information Provider(s) and, where possible, provide a link to this licence.	Contains Natural Resources Wales information © Natural Resources Wales and Database Right. All rights Reserved. Contains Ordnance Survey Data. Ordnance Survey Licence number 100019741. Crown Copyright and Database Right.

Chapter	Data	License	Source	Restrictions	Copyright Statement
7	Northern Ireland Sites of Special Scientific Interest (SSSI), Special Areas of Conservation (SACs) and Special Protection Areas (SPAs)	Open Government Licence (OGL) v3.0 https://www.nationalarchives.gov.uk/doc/open-government-licence/version/3	Open Data NI shapefile/API downloaded from: https://www.opendatani.gov.uk/ WMS assessed via QGIS	Acknowledge the source of the Information in your product or application by including or linking to any attribution statement specified by the Information Provider(s) and, where possible, provide a link to this licence.	©NIEA, 2019 [dataset name] is licensed under the Open Government Licence: http://www.nationalarchives.gov.uk/doc/open-government-licence/version/3/
8	TomBio Tools Plugin	Field Studies Council https://www.fscbiodiversity.uk/qgisplugin/nbnatlastool	Plugin downloaded from QGIS	None	Field Studies Council https://www.fscbiodiversity.uk/qgisplugin/nbnatlastool
8	Beaver_Tay_SNH.csv	Open Government Licence (OGL) v3.0 https://www.nationalarchives.gov.uk/doc/open-government-licence/version/3/	Derived from Nature Scot (Scottish Natural Heritage) spreadsheet dataset accessed via National Biodiversity Network using data from Campbell, R.D., Harrington, A., Ross, A. and Harrington, L. 2012. Distribution, population assessment and activities of beavers in Tayside. *NatureScot Commissioned Report No. 540.* https://registry.nbnatlas.org/public/show/dr333	Acknowledge the source of the Information in your product or application by including or linking to any attribution statement specified by the Information Provider(s) and, where possible, provide a link to this licence.	NatureScot (2019). Survey of distribution of beavers in Tayside in 2012. Occurrence dataset on the NBN Atlas.

Chapter	Data	License	Source	Restrictions	Copyright Statement
8	Squirrel Collation for Wales	Open Government Licence (OGL) v3.0 https://www. nationalarchives.gov.uk/doc/ open-government-licence/ version/3/	Natural Resources Wales Squirrel Collation for Wales	Acknowledge the source of the Information in your product or application by including or linking to any attribution statement specified by the Information Provider(s) and, where possible, provide a link to this licence.	Records provided by Natural Resources Wales, accessed through NBN Atlas website. Natural Resources Wales (2023) Squirrel Collation for Wales. Occurrence dataset accessed through the NBN Atlas. doi:10.15468/gh3rt3
9	NG95_line.shp; NG96_line.shp; NH05_line.shp; NH06_line.shp	Open Government Licence (OGL) v3.0 https://www. nationalarchives.gov.uk/doc/ open-government-licence/ version/3/	OS Terrain® 50 Ordnance Survey Downloaded from: https://osdatahub.os.uk/ downloads/open	Acknowledge the source of the Information in your product or application by including or linking to any attribution statement specified by the Information Provider(s) and, where possible, provide a link to this licence.	Contains OS data © Crown copyright and database right (2023)
9	Water_Voles_Beinn_ Eighe.gpx	Open Government Licence (OGL) v3.0 https://www. nationalarchives.gov.uk/doc/ open-government-licence/ version/3/	Derived from Nature Scot (Scottish Natural Heritage) spreadsheet dataset accessed via National Biodiversity Network using data from Water vole survey of Beinn Eighe National Nature Reserve - October 2011 https://registry.nbnatlas. org/public/show/dr344	Acknowledge the source of the Information in your product or application by including or linking to any attribution statement specified by the Information Provider(s) and, where possible, provide a link to this licence.	NatureScot (2019). Water vole survey of Beinn Eighe National Nature Reserve - October 2011. Occurrence dataset on the NBN Atlas

Chapter	Data	License	Source	Restrictions	Copyright Statement
10–11	Collymoon Sketch grid.jpg; Collymoon Sketch no grid; Collymoon SSSI no grid georeferenced. TIF	Esri World Imagery Map Terms of Use	Derived from World Imagery Available at: https://www.arcgis.com/home/item.html?id=8e90a00a6845a49262e0b756f57a10

WMS assessed via QGIS | The World Imagery map may be used in various ArcGIS applications to support data collection and editing. Esri and its imagery contributors grant Users the non-exclusive right to use the World Imagery map to trace features and validate edits in the creation of vector data. Users that create vector data from the World Imagery map may want to publicly share that vector data through a GIS data clearinghouse of its own or through another open data site. This public sharing could be achieved through ArcGIS Open Data or the OpenStreetMap (OSM) Initiative. For ArcGIS users that want to contribute such vector data to OSM, Esri provides applications and services directly accessible from ArcGIS platform. Users acknowledge that any vector data contributed to OSM is then governed by and released under the OpenStreetMap License (e.g. ODbL). Except for the additional limited rights granted above, any and all other uses of the World Imagery map remain subject to the terms and conditions set forth in the Esri Master Agreement or Terms of Use, as applicable. Esri and its imagery contributors retain all right, title, and interest in and to their respective imagery data contributed to the World Imagery map. | Source: Esri, Maxar, GeoEye, Earthstar Geographics, CNES/Airbus DS, USDA, USGS, AeroGRID, IGN, and the GIS User Community |

Chapter	Data	License	Source	Restrictions	Copyright Statement
11	P1 Habitat Survey Toolkit QGIS QML Line Style File with unclassified.qml; P1 Habitat Survey Toolkit QGIS QML Polygon Style File.qml; P1 Habitat Survey Toolkit QGIS QML Target Note Style File.qml	Field Studies Council compiled Phase 1 Habitat style files	QGIS mapping styles for UK habitats downloaded from: https://www.fscbiodiversity.uk/qgis-mapping-styles-uk-habitats	None	QGIS mapping styles for UK habitats downloaded from: https://www.fscbiodiversity.uk/qgis-mapping-styles-uk-habitats

Chapter	Screenshots	License	Source	Restrictions	Copyright Statement
4	1.1, 1.2	Creative Commons Attribution-ShareAlike 3.0 licence (CC BY-SA) https://qgis.org/en/site/	QGIS website https://qgis.org/en/site/	**Attribution** — You must give appropriate credit, provide a link to the license, and indicate if changes were made. You may do so in any reasonable manner, but not in any way that suggests the licensor endorses you or your use.	QGIS Development Team (2023). QGIS Geographic Information System. Open Source Geospatial Foundation Project. http://qgis.osgeo.org
7	4.25–4.28	Open Government Licence (OGL) v3.0 https://www.nationalarchives.gov.uk/doc/open-government-licence/version/3/	Nature Scot (Scottish Natural Heritage) https://opendata.nature.scot/	Acknowledge the source of the Information in your product or application by including or linking to any attribution statement specified by the Information Provider(s) and, where possible, provide a link to this licence.	© Nature Scot

Chapter	Screenshots	License	Source	Restrictions	Copyright Statement
7	4.34–4.37	Open Government Licence (OGL) v3.0 https://www.nationalarchives.gov.uk/doc/open-government-licence/version/3/	Natural England https://naturalengland-defra.opendata.arcgis.com/	Acknowledge the source of the Information in your product or application by including or linking to any attribution statement specified by the Information Provider(s) and, where possible, provide a link to this licence.	© Natural England copyright
7	4.40–4.43	Open Government Licence (OGL) v3.0 https://www.nationalarchives.gov.uk/doc/open-government-licence/version/3/	Data Map Wales https://datamap.gov.wales/	Acknowledge the source of the Information in your product or application by including or linking to any attribution statement specified by the Information Provider(s) and, where possible, provide a link to this licence.	© Natural Resources Wales
7	4.45–4.48	Open Government Licence (OGL) v3.0 https://www.nationalarchives.gov.uk/doc/open-government-licence/version/3/	Open Data NI https://www.opendatani.gov.uk/	Acknowledge the source of the Information in your product or application by including or linking to any attribution statement specified by the Information Provider(s) and, where possible, provide a link to this licence.	©NIEA

Further reading

1. QGIS Manuals For not or each version available at: https://qgis.org/en/docs/index.html:
    Desktop User Guide/Manual
    Server Guide/Manual
    QGIS Training Manual
    A Gentle Introduction to GIS
    PyQGIS cookbook
2. *An Introduction to Spatial Data Analysis: Remote Sensing and GIS with Open Source Software*
    by Martin Wegmann, Jakob Schwalb-Willmann and Stefan Dech. Pelagic Publishing,
    2020.
3. *Remote Sensing and GIS for Ecologists: Using Open Source Software.*
    Edited by Martin Wegmann, Benjamin Leutner and Stefan Dech. Pelagic Publishing,
    2016.

Index

Printed in the USA
CPSIA information can be obtained
at www.ICGtesting.com
JSHW071509100924
69474JS00002B/8